Oxford Chemistry Series

General Editors
P. W. ATKINS   J. S. E. HOLKER   A. K. HOLLIDAY

Oxford Chemistry Series

D. J. SPEDDING

Air pollution

Clarendon Press · Oxford · 1974

CHEMISTRY

Oxford University Press, Ely House, London W.1

GLASGOW NEW YORK TORONTO MELBOURNE WELLINGTON
CAPE TOWN IBADAN NAIROBI DAR ES SALAAM LUSAKA ADDIS ABABA
DELHI BOMBAY CALCUTTA MADRAS KARACHI LAHORE DACCA
KUALA LUMPUR SINGAPORE HONG KONG TOKYO

CASEBOUND ISBN 0 19 855463 X
PAPERBACK ISBN 0 19 855464 8

© OXFORD UNIVERSITY PRESS 1974

15-313406

PRINTED IN GREAT BRITAIN BY
J. W. ARROWSMITH LTD., BRISTOL, ENGLAND

Editor's Foreword

CHEMISTRY and pollution mean the same thing for many people, and the chemist is almost automatically regarded as a culpable polluter. After reading this book, the chemistry student should have no difficulty in correcting this view, and in shifting much of the culpability from man to nature. More importantly, he should see the relevance of much laboratory chemistry to the problems of atmospheric pollution, and recognize that the chemist has much to offer in the search for an atmosphere which is more congenial, in both the long and the short term, to man.

For more information about the atmosphere the reader should consult W. S. Fyfe: *Geochemistry* (OCS 16), while further discussion of motor-vehicle emissions and their control will be found in G. C. Bond: *Heterogeneous catalysis: principles and applications* (OCS 18).

A.K.H.

Acknowledgments

THE sources of data that have been published elsewhere are indicated in the text. Acknowledgements are gratefully made to the authors for permission to use the data, and to the following publishers for their permission to reproduce material from their publications.

Academic Press (*Photochemistry of air pollution*, by P. A. Leighton), Air Pollution Control Association (*Journal of the Air Pollution Control Association*), American Chemical Society (*Environmental Science and Technology*), American Geophysics Union (*Journal of Geophysics Research*), British Joint Corrosion Group (*British Corrosion Journal*), and the Clean Air Society of Australia and New Zealand (*Proceedings of the International Clean Air Congress*, Melbourne, 1972).

I wish also to express my sincere appreciation to Dr. P. Brimblecombe, Dr. J. R. Duncan and Mr. R. W. Meadows for their helpful comments, criticisms, and discussions during the preparation of the manuscript; to Professor Holliday for his very pertinent and helpful editorial advice; and to Miss S. Groves for her very competent typing of the manuscript.

Finally I wish to acknowledge the encouragement and enthusiasm of my wife, without which the manuscript would not yet be completed.

Contents

Introduction

AIR pollution studies are of a multi-disciplinary nature embodying subjects as widespread as sociology and physics, and law and botany. In this book an attempt has been made to cover the major aspects of air pollution emphasizing the discipline of chemistry. The material from other disciplines that appears in the text is not treated to the same depth as is that of chemistry. It is to be hoped that the material outside chemistry is sufficient to broaden the view of the reader specializing in chemistry and to stimulate him to read more widely on other aspects of air pollution.

Throughout the book it has been assumed that an air pollutant is some form of material added to the atmosphere as the result of the activity of man. Also an attempt has been made to show that, in almost all cases, materials regarded as atmospheric pollutants have a natural occurrence in the atmosphere. With this in mind it is hoped that the reader will emerge with a more balanced view of atmospheric pollution than that which prevails in the popular press.

Some attempt has also been made to reveal the poor state of knowledge of air pollution chemistry that currently prevails. Where conflicting theories exist, both have been given, so that the reader may assess which is the more likely. It is my hope that at least some of the readers of this book will be sufficiently stimulated by the breadth of research that remains to be done, that they will orient their work to the field of air pollution.

1. The atmosphere

THE earth's atmosphere is an envelope of gases extending to a height of about 2000 km. The density of these gases decreases with increasing altitude to such an extent that one half of the total mass of the atmosphere is found in the lower 5 km. Temperature also varies with altitude and this is used to divide the atmosphere into layers (Fig. 1.1). The properties of each of these layers have some relevance to air pollution chemistry although the properties of the troposphere are of the greatest significance since it contains the air that we breathe and the air in which all weather processes occur.

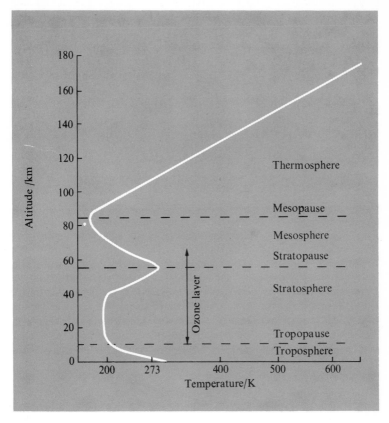

FIG.1.1. Temperature profile of the atmosphere.

The troposphere

The troposphere is characterized by a steady decrease of temperature with altitude averaging 0·6°C per 100 m. It is maintained as a relatively distinct layer of the atmosphere by the cooler air of the stratosphere which lies above it. The portion of the atmosphere at which the negative temperature gradient of the troposphere changes to constant temperature is known as the tropopause. The tropopause is not continuous but generally has two and sometimes three separate levels of discontinuity at different latitudes in both hemispheres. The tropical tropopause has an altitude of about 16–18 km. There is a distinct break at 35–50° in both hemispheres between this layer and the polar tropopause which has an altitude of 9–10 km. A third tropopause, the mid-latitude tropopause, is found between the polar and tropical tropopauses and has an altitude of 10–12 km. The positions of the tropopauses oscillate in altitude and latitude from day to day and from season to season. Even during one 24 hour period the height of the tropopause above a given point may vary by several kilometers.

There is a relatively slow rate of exchange of material through the tropopause in either direction. Most of the exchange from the troposphere occurs at the tropical tropopause, while exchange to the troposphere occurs mostly in the mid-latitude tropopause. The lifetime for exchange from the stratosphere to the troposphere is of the order of months. This exchange rate is of obvious importance for pollutants injected into the stratosphere and which are destroyed in the troposphere or at the surface of the earth. Included in this category are some of the effluents of aircraft and the debris from atmospheric nuclear explosions.

Within the troposphere the mixing time of a given hemisphere is of the order of weeks, while complete exchange between hemispheres requires about a year. Much of the mass of gaseous air pollutants is emitted in the Northern Hemisphere, hence the slow inter-hemispheric mixing time is of importance when considering the global consequences of an atmospheric pollutant.

The stratosphere

The physical properties of the stratosphere are similar to those of the troposphere except that there is a reversal in the temperature gradient, with the temperature rising to 10–20°C at 60 km. Mixing within the layer is quite marked with strong horizontal air currents and considerable vertical mixing. Very little water vapour is found in the stratosphere thus processes associated with precipitation do not occur.

The mesosphere

The temperature of this layer falls again with increase in altitude reaching −70°C at the mesopause. The rise in temperature within the stratosphere is thought to be associated with the absorption of ultra-violet and infrared radiation from the sun by ozone (see Chapter 4). The ozone concentration in the mesosphere decreases rapidly with height, hence the temperature decrease is probably due to decreased absorption of solar radiation by ozone.

The thermosphere

This is also known as the ionosphere and is the highest layer yet recognized. It is characterized by a steady rise in temperature with altitude. The temperature at 200 km exceeds 500°C and at the upper boundary (700–800 km) it exceeds 1000°C. The temperature increase is related to the absorption of solar ultraviolet radiation by molecular oxygen and nitrogen.

The air in the thermosphere becomes ionized under the influence of solar radiation. The ionized particles are found in a number of sublayers in the thermosphere and are responsible for reflecting radio waves.

Composition of the atmosphere

Air is a rather stable mixture of gases whose relative proportions vary by no more than a few thousandths of one per cent near the surface (Table 1.1). There are some exceptions to this, the most important being water vapour which is confined almost exclusively to the troposphere by the processes of condensation and precipitation. Within the troposphere water vapour can reach up to 4 per cent by volume at some points, while at other points it can be almost totally absent.

TABLE 1.1

Composition of dry air at sea level

Component	Volume per cent
Nitrogen	78·084
Oxygen	20·946
Argon	0·934
Carbon dioxide†	0·321
Neon	0·001 82
Helium	0·000 52
Krypton	0·000 11
Xenon	0·000 008 7
Methane	0·000 125

† Value obtained in Antarctica in 1971.

Variable trace amounts of other gases are found in the atmosphere. These include various hydrocarbons, carbon monoxide, nitrogen oxides, hydrogen, ammonia, hydrogen peroxide, halogens, radon, sulphur dioxide, hydrogen sulphide, organic sulphides, and mercaptans.

Ozone shows a remarkable variation throughout the atmosphere, being found largely in the stratosphere. This is due to a combination of photochemical production and destruction reactions, together with the considerable reactivity of ozone toward other atmospheric components (see Chapter 4).

The relative proportions of the major permanent constituents remain almost unchanged up to at least 80 km. Above this point the production of atomic oxygen and atomic nitrogen becomes significant, thus changing the volume percentage composition of the two major atmospheric constituents.

It is of interest to note that all of the gases mentioned in this chapter can be produced by natural sources rather than by man's activity. These gases can hence be considered to be components of the unpolluted atmosphere.

2. Aerosols

AN aerosol in the atmosphere is defined as being a dispersion of solid or liquid matter in air. In this chapter it will be shown that, when the natural aerosol is supplemented by pollutant aerosol in the form of smoke, ash, acid mists, and soluble salts, the properties of the atmosphere may be adversely affected.

Processes of absorption and scattering of light by smoke and $(NH_4)_2SO_4$ aerosols greatly affect visibility. The absorption of gases by the solid aerosol provides a means of transmitting soluble toxic gases deeper into human lungs than would be the case for the gas alone. The deposition of solid aerosols, especially soot, on exterior surfaces of buildings is a very obvious, and expensive, form of pollution by solid aerosols. The deposition of acid mists, especially H_2SO_4 mist leads to accelerated corrosion of metals while mists of carcinogenic hydrocarbons from combustion processes have obvious effects on human health.

Particle size

The particle size of the aerosol is controlled by physical processes. The upper limit is under the control of gravitational forces while the lower limit is controlled by coagulation processes. Table 2.1 lists some of the nomenclature which apply to the atmospheric aerosol. It can be seen that particle size is of basic importance. Electron microscopy shows that solid particles in the atmosphere vary markedly in configuration, ranging from almost spherical fumes to dusts of very irregular shape. The usual method of expressing particle size is the Stokes radius which is defined as the radius of a sphere having the same falling velocity as the particle, and a density equal to that of

TABLE 2.1

Terms used to describe the atmospheric aerosol

Aitken particles	Particles of less than 0·1 μm radius
Large particles	Particles of radii in the range 0·1 to 1 μm.
Giant particles	Particles of radii greater than 1 μm
Dust	Solid particles broken down from solid material and dispersed by air currents.
Fume	Solid or liquid particles formed by condensation in the vapour phase
Smoke	A fume formed as the result of a combustion process.

the material in the particle. The Stokes radius of an isolated fume particle is almost identical with its geometrical radius but the Stokes radius of a dust particle formed from the coagulation of several other particles may be very much less than the measured 'radius'. Care is thus necessary in interpreting possible physical effects due to particle size, especially optical effects.

Size distribution

It is valuable to be able to express the number of particles of a given size for all the particle sizes in the atmospheric aerosol. Both particle number and particle size range over several orders of magnitude so a logarithmic representation is used (eqn 2.1).

$$n(r) = \frac{dN}{d(\log r)}\, cm^{-3} \tag{2.1}$$

where N is the total number of aerosol particles of radius smaller than r and $n(r)$ is the number of particles of radius between r and $r + \delta r$. Fig. 2.1 shows the usual graphical representation of this function. The graph is basically a curve through a histogram, hence the total number of particles of radius between r and $r + \delta r$, is represented by the area under the curve between these radius limits.

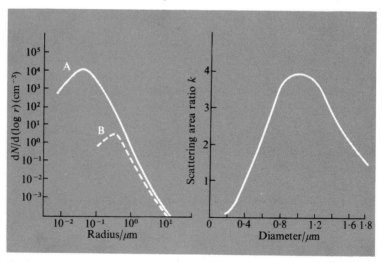

Fɪɢ. 2.1. Log number–radius distribution for a continental aerosol (A) and a maritime aerosol (B).

Fɪɢ. 2.2. Scattering area ratio as a function of particle diameter for spherical particles at wavelength 524 nm.

The log radius–number distribution for the aerosol found over large land masses (continental aerosol) shows a maximum at about $0.03 \mu m$ radius, i.e. most particles in the continental aerosol are Aitken particles. On the other hand it can be shown that the Aitken particles contribute no more than 20 per cent of the total mass of the continental aerosol. The remainder of the mass is approximately evenly divided between the large and giant particles. The aerosol found over the ocean (maritime aerosol) has a lower concentration of particles and a number maximum at greater than $0.1 \mu m$ radius in the log radius–number distribution. As sea salt forms almost all of the mass of the maritime aerosol its average particle radius varies with humidity, which controls the condensation of water vapour on the salt crystal. There is almost an order of magnitude difference between the radius of a dry sea salt crystal at low humidity and the radius of the droplet formed on this crystal at 99 per cent relative humidity.

A simplified explanation for the log radius–number distribution of the atmospheric aerosol is that it represents a balance between coagulation and sedimentation processes. The very small particles show a strong tendency to coagulate, forming particles of higher radius, while the very heavy particles are few in number due to their high sedimentation rate.

Sedimentation

When a particle falls through air a frictional drag due to the viscosity of the air is exerted on the moving particle. In order to maintain a uniform velocity, a constant force must be applied to overcome the viscous drag of the air. Stokes Law (eqn 2.2) expresses the magnitude of this force.

$$f = 6\pi r \eta u \tag{2.2}$$

where r is the radius of the small sphere, u is the velocity of the sphere and η is the coefficient of viscosity of the air. If the particle is falling under the influence of gravity the downward force F_1 is given by eqn (2.3).

$$F_1 = 4/3\pi r^3(p-p')g \tag{2.3}$$

where p is the density of the particle, p' is the density of air and g is the acceleration due to gravity. At a certain point in the fall of the particle, constant velocity will be reached, i.e. $f = F_1$ thus:

$$4/3\pi r^3(p-p')g = 6\pi r \eta u \tag{2.4}$$

or

$$u = \frac{2gr^2(p-p')}{9\eta} \tag{2.5}$$

When the particle radius approaches the mean free path of air the viscous drag of the air may be regarded as discontinuous, and some of the particles may 'slip' between air molecules. A correction factor may be applied to eqn (2.5) to account for this.

Eqn (2.5) shows that the terminal velocity u of an aerosol depends upon the square of its radius. The range of terminal velocities for the atmospheric aerosol is thus very wide e.g. at 101·3 kPa pressure and 20°C a spherical particle of density 1 g cm^{-3} and radius 0·1 μm has a terminal velocity of 9×10^{-5} cm s^{-1} while a similar particle of radius 20 μm has a terminal velocity of 1·2 cm s^{-1}.

Sedimentation controls the upper limit of atmospheric aerosol size. It is of interest to note that natural particles, such as pollens and loess (wind-blown soil) have a remarkably uniform radius of about 10 μm, which suggests that particles with terminal velocities less than those of pollens and loess (about (0·3 cm s^{-1}) remain airborne for some time.

Coagulation

Coagulation occurs when particles, rather than bouncing apart on contact, accrete or coalesce. The process is a continuous one, so that the number of particles decreases with time, while the size of the particles increases with time. Smoluchowski has derived an equation (eqn 2.6) to describe the change in concentration with time, of the number of homogeneous particles in a given volume.

$$-\frac{\mathrm{d}n_t}{\mathrm{d}t} = Kn_t^2 \qquad (2.6)$$

where n_t is the number of particles present at time t and K is a constant that depends upon particle size and diffusivity and a number of physical properties of the air. The smaller a particle the higher its diffusivity, hence very small particles have a high rate of coagulation. Particles of less than 0·01 μm radius have diffusion coefficients greater than 5×10^{-4} cm^2 s^{-1} and are rapidly coagulated. Particles larger than 0·3 μm radius have diffusion coefficients of less than 2×10^{-6} cm^2 s^{-1} which is negligibly small and thus act as acceptors on to which the rapidly diffusing smaller particles coagulate. The larger particles so formed are removed by sedimentation.

Condensation of water

The normal particle present in the atmospheric aerosol is regarded as being a mixture of soluble and insoluble materials. The soluble fraction is very important in the formation of droplets in the atmosphere. When the relative humidity of the air above a dry soluble salt exceeds the relative humidity above a saturated solution of that salt, water is condensed on to the particle containing the salt. The dry particle thus becomes a saturated salt solution containing some insoluble material. As the relative humidity in the air increases, so the radius of the droplet increases and the solution becomes less saturated. As we have seen in the previous section, the larger the particle the greater the chance of deposition by sedimentation. The probability of rainfall at high relative humidity is thus related to the condensation of water vapour on the atmospheric aerosol.

In the maritime aerosol, NaCl is an important soluble salt. At a relative humidity of 75 per cent at 20°C a dry crystal of NaCl becomes a saturated solution. In the continental aerosol, $(NH_4)_2SO_4$ is an important soluble salt. At a relative humidity of 81 per cent at 20°C a dry crystal of $(NH_4)_2SO_4$ becomes a saturated solution. The predominant salt in the atmospheric aerosol therefore has an influence on the formation of droplets at a given relative humidity.

It can be shown that the large and giant particles are very much more important in droplet formation than are the Aitken particles. The reason for this can be seen from eqn (2.7).

$$P/P_0 = 1 + \frac{c_1}{r} - \frac{c_2}{(r^3 - r_0^3)} \qquad (2.7)$$

where P/P_0 is the relative humidity in equilibrium with the droplet, r is the radius of the droplet, r_0 is the radius of the insoluble matter and c_1 and c_2 are constants. The term c_1/r accounts for the increase in relative humidity due to the Thomson effect of curvature. For radii above $0.1\ \mu m$ this is smaller than 1 per cent relative humidity but it increases with decreasing radius reaching about 500 per cent relative humidity at a radius of $10^{-3}\ \mu m$. Above $0.1\ \mu m$, therefore, the relative humidity is controlled by the last term in eqn (2.7). This term expresses the depression in water vapour pressure due to the concentration of salt in solution, i.e. Raoult's Law.

Most of the mass of atmospheric aerosol resides in the large and giant particles. These particles are also the most important in condensing water vapour and thus producing rain. Much of the mass of aerosol in the atmosphere is thus removed by washout in rain.

Effects on visibility

The most obvious effect of atmospheric pollution is a reduction in visibility. When visibility is reported in meteorological data, it refers to a measurement of the greatest distance at which a dark object of reasonable size may be seen against the horizon sky. The ability to see such an object depends upon the transmissions of light through the atmosphere, and on the contrast of the object to the background. Both of these factors are influenced by absorption and scattering processes in the atmosphere.

If a parallel beam of light is transmitted through a uniform atmosphere its intensity I falls exponentially with distance x (see eqn 2.8).

$$I = I_0 e^{-\sigma x} \qquad (2.8)$$

Here I_0 is the original intensity and σ is the extinction coefficient. The extinction coefficient may be expressed as the sum of scattering effects (b) and absorption (k) as in eqn (2.9).

$$\sigma = b + k \qquad (2.9)$$

Atmospheric gases may have an effect on the extinction coefficient by both scattering and absorption. Scattering occurs mainly as Rayleigh scattering, where the frequency of the incident light is unchanged in the scattering process. This occurs with gas molecules and particles of size much smaller than the wavelength of the incident light i.e. less than $0.1\ \mu m$ radius. Aitken particles are thus of importance in this phenomenon. The contribution of Rayleigh scattering in the reduction of atmospheric visibility is, however, very small compared with other effects.

The gases normally present in the atmosphere do not absorb visible light and, of the pollutant gases, only NO_2 is present in sufficient concentration to have any significant effect. It can be calculated that, in an atmosphere with a visual range of 16 km in the absence of NO_2, a concentration of $0.5\ mg\ kg^{-1}$ NO_2 would produce a significant discolouration of the atmosphere. The discolouration is yellowish-brown, as NO_2 absorbs significantly in the blue–green portion of the visible spectrum.

The scattering caused by the atmospheric aerosol is due to particles of a size comparable to the wavelength of visible light (0.4–$0.8\ \mu m$) i.e. large particles. This type of scattering is known as Mie scattering, and the scattering coefficient b in this case may be obtained from eqn (2.10).

$$b = NK\pi r^2 \qquad (2.10)$$

where N is the number of particles of radius r, and K is the scattering area ratio for these particles. This ratio is dependent upon the particle radius and refractive index, and on the wavelength of the incident light (see Fig. 2.2, see p. 8).

The absorption of light by the atmospheric aerosol appears to have an effect comparable to the scattering of light by Mie scattering. The absorption effect is, of course, related to the colour of the aerosol particles.

The preceding discussion relates to the reduction in light intensity due to the components of the atmosphere. The meteorological definition of visibility relates to the contrast of an object relative to the background. The contrast is affected by the processes influencing light transmission and it can be shown that the apparent contrast C at a distance x can be determined from eqn (2.11).

$$C = C_0\,e^{-\sigma x} \qquad (2.11)$$

where C_0 is the actual contrast of the object and σ is the extinction coefficient. In the definition of visibility the target is assumed to be black, i.e. $C_0 = -1$, hence eqn (8.11) may be written (eqn 2.12),

$$-C = e^{-\sigma x}. \qquad (2.12)$$

The eye is used as the sensor in the determination of visibility and it is assumed to have the ability to determine a contrast down to a limit of 2 per cent i.e. $C = 0.02$. If this value is applied to eqn (2.12) the distance x becomes visual range V, by the definition of visual range (eqn 2.13),

$$e^{-\sigma x} = 0.02. \tag{2.13}$$

From this V is found to have the value given in eqn (2.14),

$$V = \frac{3.9}{\sigma}. \tag{2.14}$$

An example of the use of this concept is illustrated in Table 2.2. At a constant ammonium ion concentration the value of σ, measured with an integrating nephelometer, falls with decreasing humidity. The visual range thus increases with decreasing humidity. We have earlier seen that the relative humidity of the air above a soluble salt influences the radius of the droplet formed by water condensing on the salt (or solution). It can also be seen (e.g. eqn 2.10), that the radius of a particle influences the value of σ. The data in Table 2.2 thus illustrate the importance of soluble salts in the atmospheric aerosol, in determining the visibility of the atmosphere (see also the section on ammonium sulphate, Chapter 8).

TABLE 2.2

Relationship between σ and particulate NH_4^+ concentrations at Stockton, England

Ammonium concentration ($\mu g\ m^{-3}$)	Relative humidity (%)	σ (km^{-1})	V (km)
20	96–97	3.7	1.1
20	89–91	2.2	1.8
20	84–86	1.2	3.3

Chemical composition of the tropospheric aerosol

The chemical nature of the tropospheric aerosol has a significant effect on the properties of the troposphere, the most important effects being in water condensation and visibility. The tropospheric aerosol has two major sources—the ocean and the land. The maritime aerosol, to a reasonable approximation, is of similar composition to the ocean, while the continental aerosol contains materials dispersed from the surface of the earth from both natural and man-made sources. Any given aerosol is a mixture of particles from both maritime and continental sources, with the maritime aerosol predominating in mid-ocean, and the continental aerosol predominating in the centre of land masses.

Man is concerned with the aerosol deposited over the land therefore the discussion following will be largely oriented to the chemical composition of that aerosol. Attempts to determine the composition of the Aitken particles in the tropospheric aerosol, have to a large extent, been unsuccessful. The large and giant particles have, however, been analyzed. Table 2.3 sets out average results for the distribution of some soluble salts between large and giant particles.

TABLE 2.3

Distribution of soluble salts between large and giant particles in the atmosphere
$(\mu g\ m^{-3})$

	Large		Giant	
	Junge	Novakov *et al.*	Junge	Novakov *et al.*
Cl^-	0·03	—	1·2	—
Na^+	0·02	—	1·2	—
SO_3^{2-}	—	2·86	—	0·42
SO_4^{2-}	4·6	1·43	1·2	0·65
NO_3^-	0·06	0	0·7	0·14
NH_4^+	0·8	0·5	0·2	0·14
Amino N	—	1·6	—	0·25
Pyridino N	—	2·0	—	0·41

The data of Junge show that NH_4^+ and SO_4^{2-} predominate in the large particles and that they exist in a mole ratio that suggests that $(NH_4)_2SO_4$ is the salt present. This has been confirmed by electron microscopy studies, and it is now known that $(NH_4)_2SO_4$ is widely distributed in the atmosphere. The importance of atmospheric $(NH_4)_2SO_4$ will be discussed in Chapter 8.

The data of Novakov *et al.* are for an aerosol from an area subject to photochemical smog episodes (Pasadena, U.S.A.). The distribution of NH_4^+ and $(SO_4^{2-} + SO_3^{2-})$ between the large and giant particles is similar to that found by Junge but the mole ratios show an excess of $(SO_4^{2-} + SO_3^{2-})$ over NH_4^+. In this aerosol a significant portion of the sulphur anions existed as salts other than ammonium salts, or as H_2SO_4 and H_2SO_3.

The Na^+ and Cl^- ions are found largely in the giant particle size range. It has been found that particles of maritime origin tend to be in this size range (see size distribution section) and elements and compounds of oceanic origin are thus concentrated in this size range. On the other hand, particles of continental origin tend to be in the large particle size range. Particles containing amino-nitrogen and pyridino-nitrogen are found mostly in the large particle size range. It is likely that these are emitted directly to the atmosphere from motor vehicles, as petrol is known to contain compounds with these functional groups.

TABLE 2.4

Concentration of some elements and salts from districts of different geography

	Windermere ($\mu g\,kg^{-1}$)	Stockton ($\mu g\,m^{-3}$)	Pasadena ($\mu g\,m^{-3}$)
Total	20	—	101·5
Al	0·26	0·8	0·8
Br	0·027	0·09	0·6
Ca	0·52	1·3	0·99
Cl	1·75	2·8	0·07
Cu	0·026	—	0·03
Cr	0·002	0·008	—
Fe	0·23	1·7	3·2
I	0·002	—	0·006
K	—	0·3	0·32
Mg	—	0·5	1·1
Mn	0·01	0·1	0·03
Na	2·3	0·8	1·0
Pb	0·09	0·4	3·3
V	0·008	0·02	0·01
Zn	0·08	—	0·18
NO_3^-	—	2·2	9·5
NH_4^+	—	4·4	0·6
SO_4^{2-}	—	11·4	12·1

Table 2.4 lists average concentrations of elements found at three sites of distinctly different geography. Lake Windermere is a rural site in Great Britain, 30 km from the nearest heavy industry; Stockton is an industrial district in Great Britain, while Pasadena is a district in the United States known for a high incidence of photochemical smog. Motor vehicle activity at the three sites is reflected especially in the concentrations of Pb, Br, and NO_3^-. Both Pb and Br are components of 'anti-knock' fluid in petrol and appear in highest concentration in Pasadena which has a higher rate of automobile activity than has Stockton. The low values obtained for Pb and Br at rural Windermere reinforce the view that these elements are of automobile origin.

The elements Na and Cl are regarded as being good indicators of the presence of sea-salt aerosol. In sea water, the ratio of Cl:Na is about 1·8. At Windermere the ratio is slightly lower than this, but sufficiently close to suggest the presence of a considerable amount of maritime aerosol. At Stockton the ratio is very high and probably reflects the presence of industrial sources of Cl. In the Pasadena aerosol the ratio is very low, probably because of reactions of Cl with photochemical smog components. It is thus obvious that the reactivity of elements, and possible anthropogenic sources, must be considered in attempting to determine their origin in the atmospheric aerosol.

It is possible to show an anthropogenic contribution to the atmospheric aerosol by selecting elements that are not likely to be of natural origin. In Table 2.4 an example of such an element is vanadium. It can be seen that in industrialized Stockton the concentration of this element is considerably higher than in Windermere or Pasadena.

The total concentration of particulate matter shown in Table 2.4 is three or four times greater than the sum of the concentrations of the elements listed. Much of this deficit appears as non-carbonate carbon, which can form up to 45 per cent of the total particulate concentration. Carbonate carbon usually forms less than 2 per cent of the total particulate concentration and appears mainly in the giant particle size range. 90 per cent of the non-carbonate carbon is found in the large particle size range.

A wide range of organic compounds has been found in the atmospheric aerosol. These include all the straight-chain alkanes from C_{18} to C_{34}, at least thirty polycyclic hydrocarbons, and many heterocyclic compounds. The aromatic polycyclic hydrocarbons are of particular interest because a number of them are known carcinogens. The best-known of this group are the benzpyrenes which are found in the atmosphere at a concentration of about 5 ng m^{-3}. They are obtained from the high temperature combustion or pyrolysis of carbonaceous materials, particularly coal tar.

Chemical reactions of atmospheric aerosols

It is to be expected that the most important reactions of atmospheric aerosols relate to their ability to absorb gases and/or catalyze gaseous reactions. When an aerosol is at equilibrium with the gases and vapours in the atmosphere it has a specific surface area of about 2–2·5 m^2 g^{-1}. It can be shown that this is approximately one half of its total specific surface area, the other half being occupied by atmospheric gases and vapours. The high surface area is accounted for by the presence of a large number of micropores most of which have a radius of less than 10 nm. These pores absorb and condense atmospheric gases including CO_2 (150–300 μg g^{-1}) CO (10–30 μg g^{-1}), CH_4 (15–60 μg g^{-1}) and NH_3 (30–100 μg g^{-1}). It is of interest to note that SO_2 in gaseous form has not been identified on an atmospheric aerosol despite the fact that most urban aerosols contain 2–4 per cent by weight of sulphur. This suggests that adsorbed SO_2 is rapidly oxidized to sulphate on aerosol particles. The presence of acid particles in the atmosphere is well known in areas of high SO_2 concentration. The acidity of these particles is attributed to H_2SO_4 formed by SO_2 oxidation. Of course, if the particle contains alkali metals or alkaline earth metals then the acidity is neutralized and sulphate salts are formed.

Smoke and ash

These are aerosol particles arising from combustion processes which may be of natural or man-made origin. Natural combustion processes include grass, brush, and forest fires as well as volcanic activity. An average grass fire extending over one acre will produce about 10^{22} fine particles which, if uniformly distributed throughout a column of air of height 3 km and cross-sectional area 4000 m^2, would give a concentration of about 10^9 particles per cm^3. The average diameter of a smoke particle is about 0·075 μm, i.e. most are Aitken particles. Much of the smoke is made up of carbonaceous compounds, particularly tarry hydrocarbons and resins. The very small size of smoke particles enables them to penetrate buildings in the same manner as gases. Unlike gases, the smoke particles have a high 'sticking' power and are found deposited on surfaces where air turbulence is high e.g. edges of window and door frames.

The solid material set free when a fuel is completely oxidized is known as ash. The size of the particles emitted from an industrial furnace depends upon the lowest velocity of the combustion gases in the chimney. We have seen, in the discussion on sedimentation, that the terminal velocity of a particle increases with size. In order that a particle is emitted from a chimney the flue gas velocity must exceed the terminal velocity. Particles of size such that the terminal velocity is not exceeded remain in the combustion system as ash. A typical flue-gas velocity in an industrial chimney is $12\,ms^{-1}$, which is sufficient to carry particles of $\leqslant 200\,\mu$m diameter out of the chimney. The larger particles sediment close to the chimney as the flue-gas velocity decreases, while the smaller particles travel further from the chimney before sedimenting. The flue-gas velocity in a domestic open-fire chimney is rarely greater than $1·5\,ms^{-1}$. In this case the maximum size of ash particle leaving the chimney is 75 μm in diameter.

Damaging effects of aerosols

Some of the atmospheric aerosol may have a damaging effect on human health because of its chemical nature. The presence of carcinogenic hydrocarbons has already been mentioned. Elements such as Pb and As may exhibit toxic effects but only at quite high concentrations which are not normally experienced in the atmosphere. The aqueous aerosol can also be damaging to human health because of its chemical nature. Sulphuric acid mists formed during the atmospheric oxidation of SO_2 are known to cause respiratory damage to laboratory animals at concentrations of $2·5\,mg\,kg^{-1}$. Mists of oils from industrial processes and from motor vehicles do not specifically damage human health at concentrations found in polluted atmospheres.

The physical properties of the atmospheric aerosol affect human health, either by allowing penetration of the lung and causing irritation to the internal membranes, or by transporting adsorbed toxic gases and vapours deeper into the lung than they would normally travel. Many gases are soluble in the layer of water covering the mucous membranes of the respiratory tract. Some of the very soluble gases are completely removed on these surfaces before reaching the lungs. However, very small particles are transported deep into the lungs and the gases adsorbed on them may be released within the lungs (see Chapter 5).

Materials in our environment may be damaged by the larger solid particulate matter by mechanical abrasion. More important however, is the deposition of the atmospheric aerosol on materials, especially buildings. Little damage is caused to the building surfaces, but the effect is unsightly and expensive to remove. Accelerated attack by corrosive gases especially SO_2 is noted in the area about deposited particulate matter on metal surfaces (see Chapter 5). This may be due to the formation of distinct electrochemical cells on the particulate material and the metal, or it may simply be due to the ability of the particulate material to concentrate the corrosive gas at a point by absorption.

3. Carbon dioxide and water vapour

THE combustion of fossil fuels to provide a source of energy is the major means by which man pollutes the atmosphere. Almost all of the fossil fuels (coal, oil, natural gas, etc.) are carbonaceous or hydrocarbon in nature. Optimum usage of these fuels requires complete oxidation in the combustion process. Carbon dioxide and water vapour are thus, in terms of total mass emitted, the major gaseous pollutants of the atmosphere. Both gases have similar effects on climate but their atmospheric reactions are so different that they will be discussed separately.

The effect of pollutant water vapour will be shown to be very small, while the effect of carbon dioxide will be seen to be possibly of considerable consequence. The concentration of carbon dioxide in the atmosphere is still showing a steady increase. It is theoretically possible to show that this may result in an increase in the temperature of the earth—the greenhouse effect. The extent to which the greenhouse effect will change the pattern of the world's climate depends upon the numerical parameters used in a theoretical model. Parameters that predict rapid disastrous climatic effects have been used while some other parameters that predict limited climatic effects have also been used. It is probably wise to err on the side of caution and seek to limit the emission of carbon dioxide to the atmosphere and thus allow the total carbon cycle to come to equilibrium through the vast reservoirs of the deep ocean and sediments.

Water vapour

Almost all of the water vapour in the atmosphere is confined to the troposphere. The total mass of atmospheric water vapour is equivalent to a precipitation of 2·5 cm of rain over the whole of the earth's surface. The average rainfall over the earth is about 90 cm per year hence there are about thirty-six evaporation–precipitation cycles per annum. This means that the average residence time for a water molecule in the atmosphere is about ten days. This is of particular importance in air chemistry as many trace substances are removed from the troposphere by water precipitation.

Substances may be removed from the atmosphere by water precipitation in one of two ways—rainout or washout. Rainout occurs within clouds where the most important process is condensation of water vapour on the solid aerosol (see Chapter 2). Much of the total mass of the natural aerosol is removed from the atmosphere when this water precipitates. Washout occurs below the clouds and is a very efficient process for the removal of large solid aerosols. After an appreciable amount of rain, however, almost all of the solid aerosol is removed from below the cloudbase. The often dramatic improvement in visibility following a shower of rain can be explained by the washout of much of the light-scattering aerosol that makes such a large contribution to haze.

Many of the soluble gases e.g. CO_2, NH_3, SO_2, NO_2 are also removed from the atmosphere by rainout and washout. As an example, the removal of SO_2 as sulphate may be quoted. Of the total sulphate concentration in rainwater it has been calculated that 70 per cent is derived from SO_2 collected during washout and only 5 per cent from SO_2 collected during rainout.

Liquid water in the atmosphere has an average residence time of about eleven hours, which is longer than the average lifetime of clouds. It can be concluded then, that most clouds evaporate. Solids dissolved or suspended in the liquid water of clouds become solid aerosol particles when clouds evaporate, so that the presence of clouds does not necessarily indicate a rapid cleansing of the atmosphere.

The heat balance of the earth is influenced by atmospheric water, both in the liquid and in the gaseous state. Only 47 per cent of the total solar radiation reaching the thermosphere is received by the surface of the earth. Liquid water in the form of clouds reflects about 20 per cent of the incoming radiation. The surface of the earth reflects about 14 per cent while the remainder is absorbed by ozone in the stratosphere, and carbon dioxide and water vapour in the troposphere.

In Fig. 3.1 it can be seen that the black-body spectrum for solar radiation is largely in the ultraviolet–visible region of the spectrum. Also shown are the absorption spectra of O_3, O_2, NO_2, H_2O and CO_2 in the same wavelength region. Ozone shows a strong absorption in the ultraviolet region and it is this that shields the earth's surface from much of the ultraviolet radiation of the sun. Little absorption can be found in the visible region while in the infrared the predominant absorption is by water vapour, which absorbs 10–20 per cent of solar radiation.

Terrestrial radiation from earth to space is due to black-body radiation at the temperature of the earth's surface i.e. about 10°C. This radiation spectrum is also shown in Fig. 3.1. Terrestrial radiation is infrared radiation and is strongly absorbed by both CO_2 and water vapour. Only radiation of 7000 nm to 14 000 nm can escape directly to space through a 'window' of low absorption in the water-vapour spectrum. Terrestrial radiation absorbed by CO_2 and water vapour increases the temperature of these gases. This energy is eventually re-emitted with a considerable portion being directed back to the earth's surface. Radiation reflected in this manner serves to keep the temperature at the surface of the earth at a higher level than would be found if the gases were absent. This effect is called the 'Greenhouse Effect', because of the obvious analogy with the warming effect of a greenhouse.

Any increase in the atmospheric water vapour concentration will necessarily increase the temperature at the earth's surface because of the greenhouse

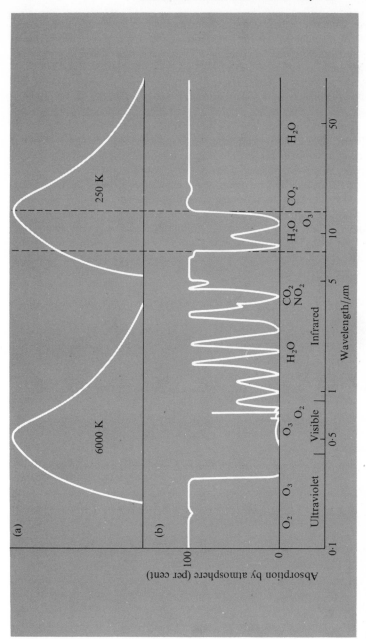

Fig. 3.1. (a) Black-body radiation spectra for sun (6000 K) and (not on same scale) earth (250 K). (b) Atmospheric absorption spectrum produced by principal absorbing gases (after Sawyer (1972)).

effect. Man's activities, however, produce only a very small portion of the total water in the water cycle. The atmospheric water vapour content is thus not measurably affected, leaving us with the conclusion that water vapour is a very innocuous pollutant gas.

Carbon dioxide

All important processes involving atmospheric CO_2 occur at the earth's surfaces. These processes may be summarized in the carbon cycle illustrated in Fig. 3.2. The atmosphere acts as a passive buffer reservoir with a large but limited capacity. The ocean is a reservoir sixty times as large as the atmosphere and is made up of two distinct layers. The surface layer of the ocean is about 100 m deep and is in a much more rapid equilibrium with the atmosphere than the deep layer which forms the bulk of the ocean. By far the largest portion of carbon in the carbon cycle is found in the marine and terrestrial sediments. The turnover of carbon within this reservoir is very slow indeed e.g. marine carbon is estimated to turn over once in 100 000 years. Man interferes with this cycle by using carbonaceous fossil fuels as an energy source and thus increases the turnover of sedimentary carbon by the production of CO_2 which is emitted into the atmosphere. The effect of this activity will be discussed later in this chapter.

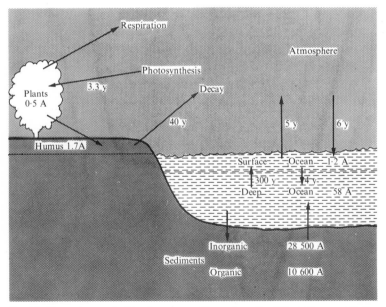

Fig 3.2. The carbon cycle. The times are residence times for CO_2 with respect to the indicated reservoir. $A = 2.5 \times 10^{18}$ g of CO_2 = atmospheric CO_2 reservoir.

In general there is no real CO_2 balance at any given point on the earth's surface. Plant life, in particular, affects the CO_2 concentration near the earth's surface by photosynthesis and respiration. Photosynthetic assimilation of CO_2 ceases at night leading to higher CO_2 concentrations at night. During the day the reverse effect may be found e.g. CO_2 concentrations below potato leaves in the field can be up to 30 per cent higher than those above the leaves. Such fluctuations may still be found 1 and 2 km into the troposphere.

One of the important steps in the carbon cycle is the exchange of CO_2 between the atmosphere and the ocean surface layer. Gas exchange at an aqueous surface is of considerable importance in atmospheric chemistry so it will be dealt with in some detail here.

At equilibrium, the mass of a gas that will dissolve in an aqueous solution is directly related to the partial pressure of that gas in the gas phase. The higher the gas-phase partial pressure, the higher the partial pressure of the gas in solution. This relationship is summarized in Henry's Law (eqn 3.1);

$$P = HX, \tag{3.1}$$

where P = partial pressure of the gas expressed as a mole fraction,

X = concentration of the gas in solution expressed as a mole fraction,

H = Henry's Law constant.

This law applies to an equilibrium situation, hence over a long time scale the amount of a given gas in solution in natural water must be in diffusion equilibrium with its partial pressure in the atmosphere. However, at any given instant, the partial pressure of the gas in the water may be above or below that in the overlying atmosphere. There will thus be a net flux of the gas one way or the other across the interface tending to bring the system back to equilibrium. To a reasonable approximation the atmosphere can be considered to be of constant composition and nearly constant pressure. The direction of movement of a gas across the air–water interface therefore depends mostly upon factors that change its partial pressure in the surface waters. These factors include seasonal water temperature variations (the solubility of CO_2 increases with decreasing temperature), and photosynthesis and respiration of aquatic plants.

When such a partial pressure difference exists the kinetics of the exchange of the gas between the two phases must be considered. It is usual to discuss these kinetics in terms of an 'exchange constant' k which is defined as follows:

$$\frac{dq}{dt} = kA(X_1 - X_2) \tag{3.2}$$

where dq/dt = rate of change of mass q of CO_2 with respect to time t,

A = area of the interface across which exchange takes place.

X_1 and X_2 represent the CO_2 concentrations in mass per unit volume in each phase.

k thus has the dimension of length time^{-1}.

With a natural water surface there is approximately a constant surface area so that the rate of net mass exchange of gas depends upon the partial pressure difference between the two phases and upon k. The physical significance of k may be described by the liquid-film model. In this model it is assumed that in both the liquid and gas phases there is a thin film of fluid where there is locally laminar flow while the bulk of the fluid is in turbulent flow (Fig. 3.3).

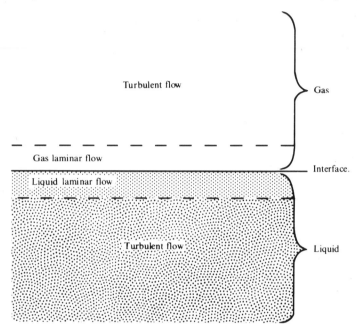

Turbulent flow

Gas

Gas laminar flow

Interface.

Liquid laminar flow

Turbulent flow

Liquid

Fig. 3.3. Diagram to illustrate the laminar-layer model as applied to a gas–liquid interface.

Gas exchange between the two phases is thought to be controlled by the molecular diffusion of the gas through both laminar layers. In the case of CO_2 the diffusion in the liquid phase is about ten thousand times slower than in the gas phase and is thus rate-determining. It is also possible to relate k for CO_2 with the molecular diffusivity D of CO_2 in the liquid and the thickness of the liquid laminar layer δ:

$$k = \frac{D}{\delta}. \tag{3.3}$$

We can see now that the rate of exchange of atmospheric CO_2 with the ocean surface depends upon the CO_2 partial pressure difference between the phases, and on the thickness of the laminar layer of the ocean. This latter varies from 10–200 μm and is determined by the wind velocity immediately above the surface. The shear stress of the wind tends to tear the laminar layer film from the bulk of the water, thus high wind velocities lead to low values of δ. Because ocean conditions vary from glassy calm where δ is at its maximum value, to violent storm where δ is very small, the rate of exchange of CO_2 must vary with weather conditions. Wind tunnel measurements have shown that CO_2 exchange is roughly proportional to the square of the wind velocity e.g. at 10 ms^{-1} wind speed, $\delta = 50 \mu$m and CO_2 exchange is twenty times greater than at 2 ms^{-1} where $\delta = 200 \mu$m.

The above discussion has assumed no chemical interaction of CO_2 with water. To obtain a full picture of CO_2 exchange the following reactions should be considered:

$$(CO_2)_g \rightleftharpoons (CO_2)_{aq} \tag{3.4}$$

$$(CO_2)_{aq} + H_2O \rightleftharpoons H_2CO_3 \tag{3.5}$$

$$H_2CO_3 \rightleftharpoons H^+ + HCO_3^- \quad K_1 = 4.47 \times 10^{-4} \text{ mol l}^{-1} \tag{3.6}$$

$$HCO_3^- \rightleftharpoons H^+ + CO_3^{2-} \quad K_2 = 5.62 \times 10^{-11} \text{ mol l}^{-1} \tag{3.7}$$

Almost all of the CO_2 in solution remains as $(CO_2)_{aq}$ with about 1 per cent as the unstable H_2CO_3. When this is taken into account the true value of K_1 is about $3 \times 10^{-4} \text{ mol l}^{-1}$. In solutions of pH less than 4 the concentration of ionic species is negligible. Sea water, however, has a pH of about 8 so ionic species must be considered together with the rates of the equilibrium reactions that produce them. CO_2 exchange is influenced by the presence of ionic carbonate species as these have a higher mobility in solution than $(CO_2)_{aq}$. As the rate of CO_2 exchange is determined by the rate of transport of $(CO_2)_{aq}$, HCO_3^-, and CO_3^{2-} across the liquid laminar layer, the higher the concentration of ionic species the greater the rate of exchange. At a fixed pH and temperature the relative proportions of $(CO_2)_{aq}$, HCO_3^-, and CO_3^{2-} are fixed by equilibria (3.5), (3.6), and (3.7). Under these conditions the rate of CO_2 exchange depends upon the thickness δ of the liquid laminar layer. Fig. 3.4 shows clearly that the difference in mass of CO_2 exchanged per unit time for ionic and non-ionic species is small at low surface layer thickness but becomes significant as the surface layer thickness increases. This difference in rate of exchange may be expressed in terms of an exchange enhancement, α, where:

$$\alpha = \frac{\text{Exchange rate taking ionic equilibria into account}}{\text{Exchange rate assuming only } (CO_2)_{aq} \text{ participating}}. \tag{3.8}$$

With seawater, α is 1·62 for a surface layer thickness of 300 μm and 2·67 for a surface layer thickness of 600 μm. Under average conditions the open ocean has a surface layer thickness of less than 100 μm so this enhancement of CO_2 exchange is of little global consequence. This should be compared with the exchange of SO_2 into natural waters (see Chapter 5).

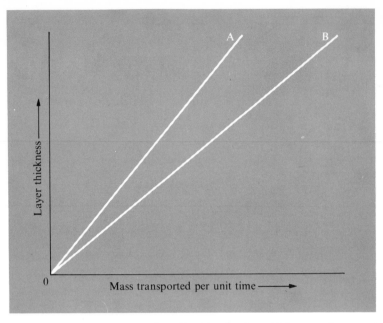

Fig. 3.4. Graph to illustrate the effect of liquid surface layer thickness on the mass of non-ionic (A) and ionic (B) CO_2 species transported in solution in unit time.

Carbon dioxide from combustion

Data which have been obtained for a number of sampling points remote from centres of man's activities indicate a constant increase in the atmospheric CO_2 concentration of about $1.0\ mg\,kg^{-1}\,y^{-1}$ (Fig. 3.5). A simple calculation of the annual mass of CO_2 emitted by combustion shows that the change in CO_2 concentration in the atmosphere due to combustion represents about one half of the emitted fossil fuel CO_2. The size of the oceanic reservoir and the solubility of CO_2 in water suggests that fossil fuel CO_2 should be removed more rapidly from the atmosphere. One of the reasons why this is not observed is that the rate of exchange of CO_2 into the large oceanic reservoir is relatively slow. Another is that the ocean is essentially a carbonate–bicarbonate buffer system and as a consequence a large increase in the atmospheric CO_2 partial pressure is necessary for a relatively small increase in the oceanic CO_2 concentration e.g. a 10 per cent increase in atmospheric CO_2 partial pressure results in a 0.6 per cent increase in oceanic CO_2 concentration.

Two major effects of increasing atmospheric CO_2 concentration are possible. Firstly, the increased CO_2 may affect photosynthesis; however, growth-chamber studies have shown that the present CO_2 concentrations

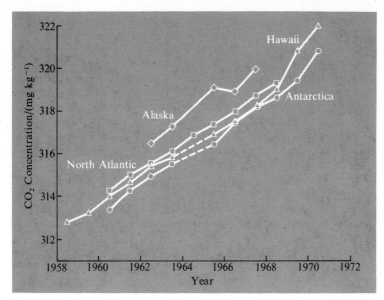

FIG. 3.5. Annual mean CO_2 concentration recorded at ◇, Alaska; □, North Atlantic; △, Hawaii; ○, Antarctica, (after Garatt and Pearman 1972).

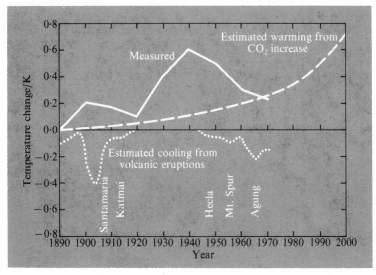

FIG. 3.6. Average Northern Hemisphere temperature changes together with estimated heating effects due to CO_2 increase and cooling due to volcanic dust (after Dyer 1972).

are below optimum for photosynthesis when other factors such as nutrient supply are optimum. Secondly, an effect on world climate may occur because of the greenhouse effect. Fig. 3.6 shows the calculated change in world temperature up to the year 2000 based on calculated emissions of CO_2 together with data on observed temperature changes during the 20th century. It can be seen that, to the present date, predicted temperature changes are smaller than natural changes so that the validity of the original predictions is in doubt. Some workers have suggested that the observed temperature increase from 1920 to 1940 was due to a CO_2 greenhouse effect which was, in 1940, decreased by increasing atmospheric turbidity from the solid particulate matter from combustion. The stratospheric aerosol from this source reflects radiant energy from the sun causing a decrease in atmospheric temperature. It is difficult to predict which of the two parameters (CO_2 greenhouse effect or atmospheric turbidity) will dominate in the future. Certainly, depending upon the parameters chosen it is possible to predict that in the coming years we will either 'freeze' or 'fry'.

4. Ozone

ALL of the important processes associated with ozone occur within the atmosphere, especially in the stratosphere. In earlier pages it has been shown that ozone is not uniformly distributed throughout the atmosphere but is confined largely to the stratosphere (Fig. 1.1). It has also been pointed out that the increase in temperature in the stratosphere is due to the absorption of solar ultraviolet radiation by ozone and that this absorption shields the surface of the earth from much of the solar ultraviolet thus reducing its damaging effects to terrestrial organisms. Reference to Fig. 3.1 will show that both oxygen and ozone absorb solar ultraviolet radiation. The photochemical reactions that are consequent upon this absorption form the basis of a photochemical equilibrium that maintains the ozone layer.

Ozone is formed in the following reactions:

$$O_2 + hv \rightarrow O + O \qquad (\lambda < 242 \text{ nm}) \qquad (4.1)$$

$$O_2 + O + M \rightarrow O_3 + M \qquad (4.2)$$

Reaction 4.2 is a three-body collision reaction where M is usually N_2 or O_2. The destruction of ozone is controlled by:

$$O_3 + hv \rightarrow O_2 + 0 \qquad (\lambda < 1180 \text{ nm}) \qquad (4.3)$$

$$O + O_3 \rightarrow 2O_2 \qquad (4.4)$$

There is thus an apparent equilibrium in the ozone region with the concentration of ozone remaining constant, however atmospheric mixing processes upset the equilibrium by removing some of the ozone to the troposphere where it is destroyed. The ozone concentration in the ozone region may reach 10 mg kg^{-1} while at sea level in unpolluted areas it is as low as 0.01 mg kg^{-1}.

Almost all tropospheric ozone of natural origin is thought to come from the stratospheric ozone layer by exchange across the tropopause. In the troposphere ozone is chemically destroyed, primarily by contact with the earth's surface but also in clouds, and by gaseous and particulate trace substances. In the latter category are some of the reactions associated with photochemical smog which are dealt with in detail in Chapter 6.

Ozone of artificial origin is formed in the troposphere from photochemical reactions of pollutant gases that give rise to atomic oxygen (see Chapter 6). The atomic oxygen may then become involved in reaction 4.2 resulting in the formation of ozone. Concentrations of ozone up to about 0.2 mg kg^{-1} can be obtained in localized areas from these reactions. In the past it has been thought that ozone from this source would only be found in regions of high sunlight intensity. Recently, however, evidence has become available that suggests that ozone levels as high as 0·1 mg kg^{-1} may be found in rural England in the summer months.

Recent observations of ozone concentrations in unpolluted areas, and experiments on artificial atmospheres suggest an alternative natural tropospheric source of ozone similar to the artificial source discussed in the preceding paragraph. The reactions producing this ozone may be described by the following general equation:

$$NO_2 + olefin + hv \rightarrow O_3 + other\ products. \qquad (4.5)$$

Olefins occur naturally in the atmosphere, particularly as terpenes which cause the blue haze seen over forested areas, while nitrogen dioxide is produced largely by microbial action in the soil. It is most unlikely that sufficient ozone would be generated under these natural conditions to cause damage to our environment. On the other hand, pollutant ozone concentrations are sufficiently high to cause environmental damage.

Many plants are damaged by ozone concentrations as low as 0·1 mg kg^{-1}. The physiology of the damage is so well-known that plants can be used as indicators of ozone pollution by observing the nature of the lesions found on leaves of different plant species. The biochemical basis of the damage is unknown, although it is known that photosynthesis is inhibited by abnormal ozone concentrations. Some workers have attributed the inhibition of photosynthesis to a partial closing of the leaf pores (stomata) induced by ozone. The closure results in a lowering of CO_2 uptake and thus of the rate of photosynthesis.

Humans are also affected by low concentrations of ozone. The odour detection limit is 0·02 mg kg^{-1} to 0·05 mg kg^{-1}, irritation of the nose and throat occurs at 0·05 mg kg^{-1} and dryness of the upper respiratory mucosa is found at 0·1 mg kg^{-1}. Tests on rats have shown that the effect of ozone is not confined to the respiratory tract but biochemical processes are also affected e.g. it is known that ozone denatures proteins and thus enzymes. Relatively high concentrations of the gas would be needed for biochemical effects to occur to any extent, as the very reactive ozone has many surfaces on which it could be destroyed before it is absorbed into the body fluids.

The best-known example of damage to non-living materials by ozone is the deterioration of rubber in the presence of pollutant ozone and sunlight. In the Los Angeles County the deterioration of motor-vehicle tyres has been used to estimate ozone concentrations. Dyes that are susceptible to colour change under oxidizing conditions are affected by atmospheric ozone and it is suspected that asphalt is damaged by low ozone concentrations. The reader is referred to standard organic chemistry texts for the nature of the reaction between ozone and organic compounds.

Summary

Most atmospheric ozone is produced in the stratosphere and mixes slowly down to the troposphere where it is found at very low concentrations. Pollutant ozone arises almost exclusively from photochemical smog reactions (Chapter 6). Ozone concentrations from this source are often sufficient to damage vegetation, motor vehicle tyres, and asphalt and may, on occasion, cause irritation to the human respiratory tract. It is probably better to consider ozone effects on the enrivonment together with the effects of the other constituents of photochemical smog rather than in isolation.

5. Sulphur dioxide

SULPHUR is present in the atmosphere in at least three forms—SO_2, H_2S, and aerosol sulphate. On a global scale these species are involved in the sulphur cycle, a recent version of which is illustrated in Fig. 5.1. Two of the processes in this cycle may be taken as being atmospheric pollution processes. The production of SO_2 by the combustion of fossil fuels is an obvious pollution source. Less obvious is the emission of H_2S to the atmosphere by biological processes on land in coastal areas. Decomposition of organic waste products from man's activities accounts for some of this H_2S emission and illustrates clearly that water pollution (with which H_2S emission is usually associated) and air pollution are very closely related.

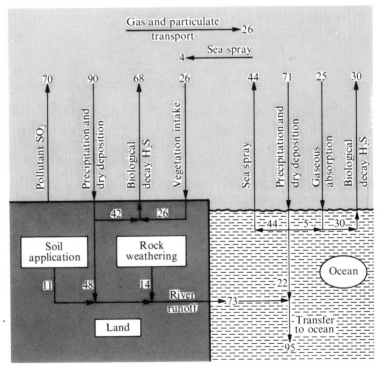

FIG. 5.1. Environmental sulphur cycle. Units 10^6 ton y^{-1} sulphur (after Robinson and Robbins 1970).

In gaseous form (SO_2 and H_2S) sulphur is quite rapidly transported in the atmosphere, but on oxidation transport is limited as sulphur is now present as solid sulphate salts or as sulphuric acid mist. Both the gaseous and the aerosol forms of atmospheric sulphur have detrimental effects on the environment. Vegetation is damaged, and some human respiratory complaints are worsened by quite low concentrations of atmospheric SO_2. Non-living materials in the environment tend to suffer damage from the acidic oxidation product of SO_2 rather than directly from the gas itself. Man especially expects his buildings and books to have a lifetime of centuries but in this time scale the amount of sulphuric acid formed on their surfaces is quite considerable. One thus finds building stone in the older industrial cities suffering from severe erosion and books stored in libraries of industrial cities deteriorating especially with respect to mechanical strength of the paper and bindings.

In this chapter the portions of the sulphur cycle that directly involve sulphur compounds emitted as pollutants to the atmosphere will be discussed, and the role of these compounds in the recent deterioration of the environment will be considered.

Sources

Table 5.1 lists the major sources of pollutant SO_2. The most prominent of these is the combustion of fossil fuels, all of which contain some sulphur compounds as 'contaminants' e.g. coal and fuel oil may have up to 3 per cent sulphur, while petrols usually contain about 0·05 per cent sulphur. Smelting of sulphide ores is the other major source.

TABLE 5.1

Annual pollutant SO_2 emissions for 1965 (after Robinson and Robbins (1970)

Source	Emission (10^6 tonnes)
Coal	102
Petroleum (combustion and refining)	28·5
Copper smelting	12·9
Lead smelting	1·5
Zinc smelting	1·3

It might be expected that such high total emissions of SO_2 would result in high atmospheric SO_2 concentrations. In localized areas near to large sources of pollution the SO_2 concentrations may reach as high as 1 mg kg^{-1} (about $3000 \mu\text{g m}^{-3}$). The background SO_2 concentration has, however, been estimated to be in the range 0·3 to 1·0 $\mu\text{g m}^{-3}$.

The magnitude of the total H_2S source is unknown. All versions of the sulphur cycle deduce the H_2S source magnitude from the mass necessary to obtain a balanced cycle. No differentiation has been made between natural H_2S sources, and sources resulting from anaerobic metabolism of organic waste compounds from man and his activities. The average tropospheric concentrations of H_2S has been estimated to be 0.3 μg m^{-3} while concentrations of some Netherlands cities have reached more than 100 μg m^{-3} probably because of reducing conditions in the canals.

Oxidation of H_2S in the atmosphere

For some time now it has been accepted that H_2S is rapidly oxidized to SO_2 in the atmosphere. Evidence for such an atmospheric oxidation is scant. Possible modes of oxidation include homogeneous gas-phase oxidation by atomic and molecular oxygen and ozone, as well as heterogeneous oxidation in fog or cloud droplets by the same compounds. Reactions involving electronically-excited H_2S are unlikely as this molecule does not absorb solar radiation of the wavelengths reaching the troposphere.

Oxidation with oxygen at atmospheric temperatures and pressures proceeds immeasurably slowly. The reaction with ozone (5.1) can be measured although it too is very slow:

$$H_2S + O_3 \rightarrow SO_2 + H_2O \qquad (5.1)$$

A more likely reaction is a chain reaction with atomic oxygen. The atomic oxygen required for this chain oxidation may be obtained from the photochemical dissociation of ozone (Chapter 4) or from photochemical smog reactions (Chapter 6).

No useful data are available on the solution oxidation of H_2S in fog or cloud droplets.

Oxidation of SO_2 in the atmosphere

One of the major steps in the sulphur cycle is the precipitation and dry deposition of sulphur from the atmosphere as sulphate in the rain, or as solid particulate matter. It is presumed that much of the atmospheric SO_2 is oxidized to sulphate and returned to the earth's surface in this form.

The rate at which SO_2 is oxidized determines its lifetime in the atmosphere provided that oxidation is the most important removal mechanism. A number of field experiments have been carried out to give the SO_2 oxidation rates summarized in Table 5.2.

The wide range of oxidation rates observed in the field imply variations in the conditions under which the measurements were made. A number of oxidation mechanisms are known and these could produce different rates under different conditions e.g. a photochemical mechanism would produce a high oxidation rate during the middle of the day while oxidation in fog droplets catalyzed by metal ions would be independent of sunlight. Most of

TABLE 5.2

Summary of SO_2 oxidation rate studies in the atmosphere (after Urone and Schroeder 1969)

Source of SO_2	Concentration of SO_2 on release from source $(mg\,kg^{-1})$	Relative humidity during measurement	Rate of SO_2 consumption in atmosphere $(\%\,ks^{-1})$
Ni smelter	0·1–1·0	—	0·6
Smelter	0·01–20·3	65–70%	190
Coal-fired power plant	2200	70–100%	1·7
Coal-fired power plant	2200	100%	8·3

the postulated mechanisms will now be discussed and their probable rates in the atmosphere compared.

Aqueous phase oxidation

A clear understanding of SO_2 oxidation in aqueous phases involves a knowledge of the solution chemistry of SO_2. The equilibria (5.2)–(5.6) outline the most important processes:

$$(SO_2)_g + H_2O \rightleftharpoons (SO_2)_{aq} \tag{5.2}$$

$$(SO_2)_{aq} + H_2O \rightleftharpoons H_2SO_3 \tag{5.3}$$

$$H_2SO_3 + H_2O \overset{K_1}{\rightleftharpoons} H_3O^+ + HSO_3^- ; \quad K_1 = 1.6 \times 10^{-2} \tag{5.4}$$

$$HSO_3^- + H_2O \overset{K_2}{\rightleftharpoons} H_3O^+ + SO_3^{2-} ; \quad K_2 = 1.0 \times 10^{-7} \tag{5.5}$$

$$2HSO_3^- \overset{K_3}{\rightleftharpoons} S_2O_5^{2-} + H_2O ; \quad K_3 = 7 \times 10^{-2}\,mol^{-1}. \tag{5.6}$$

A solution of SO_2 in water thus contains hydrated SO_2, H_2SO_3, HSO_3^-, SO_3^{2-}, and $S_2O_5^{2-}$ in proportions that vary with pH and concentration. Sulphurous acid, H_2SO_3, is unknown as a free acid and because of this it is often represented as $H_2O \cdot SO_2$. Rainwater usually has a pH of 4 to 6 thus the predominant ion in atmospheric water is HSO_3^-.

Oxidation of a solution of sulphur compounds in oxidation state $+4$ as represented by eqns (5.2) to (5.6) may be direct to oxidation state $+6$ (sulphate) or it may pass through oxidation state $+5$ (dithionate):

$$(S^{IV})_{aq} \overset{O_2}{\longrightarrow} (S^{VI})_{aq}$$
$$\overset{O_2}{\searrow} (S^V)_{aq} \overset{O_2}{\nearrow}$$

Sulphate is the only higher oxidation state sulphur compound found in atmospheric water. Dithionate has never been reported but this may be because no great effort has been made to detect it.

The direct oxidation of $(S^{IV})_{aq}$ by atmospheric oxygen proceeds very slowly, if at all. The presence of metal ion catalysts, especially Mn^{2+}, Fe^{3+}, and Cu^{2+} results in a rapid oxidation. The general oxidation pathway has recently been deduced from flash photolysis studies:

Initiation:

$$SO_3^{2-} \xrightarrow{h\nu} SO_3^- + e_{aq}^- \qquad (5.7)$$

$$SO_3^- + O_2 \rightarrow SO_5^- \qquad (5.8)$$

Propagation:

$$SO_5^- + SO_3^{2-} \rightarrow SO_4^- + SO_4^{2-} \qquad (5.9)$$

$$SO_4^- + SO_3^{2-} \rightarrow SO_4^{2-} + SO_3^- \qquad (5.10)$$

Termination:

$$SO_5^- + SO_5^- \rightarrow products \qquad (5.11)$$

$$SO_4^- + SO_4^- \rightarrow products \qquad (5.12)$$

In solutions of lower pH, where HSO_3^- predominates, reaction (5.10) becomes:

$$SO_4^- + HSO_3^- \rightarrow HSO_4^- + SO_3^- \qquad (5.13)$$

Reaction (5.13) is about 2·5 times faster than reaction (5.10) and thus explains an increased oxidation rate at a pH approximating to the value of pK_1 for reaction (5.4).

Reaction (5.7) has been given for flash photolytic initiation. In thermal autoxidation this reaction is replaced by some other electron transfer e.g.

$$SO_3^{2-} + Cu^{2+} \rightarrow SO_3^- + Cu^+ \qquad (5.14)$$

It is difficult, however, to provide a reaction similar to (5.14) for the case where Mn^{2+} acts as a catalyst, as Mn^I exists in only a few complexes in non-aqueous solvents. A mechanism involving Mn^{II}–SO_2 complexes may be used:

$$SO_2 + Mn^{2+} \rightarrow [MnSO_2]^{2+} \qquad (5.15)$$

$$2[Mn \cdot SO_2]^{2+} + O_2 \rightarrow [(Mn \cdot SO_2^{2+})_2O_2] \rightarrow 2[MnSO_3]^{2+} \qquad (5.16)$$

$$[MnSO_3]^{2+} + H_2O \rightarrow Mn^{2+} + HSO_4^- + H^+ \qquad (5.17)$$

Reaction (5.16) is probably rate-determining as Mn-bound oxygen is re-arranged to S-bound oxygen in the oxidation step.

Similar mechanisms have been postulated for other metal ion catalysts. Whether the radical chain mechanism or the metal complex mechanism or a combination of both is correct awaits further experimental elucidation.

An important observation from the point of view of atmospheric chemistry is that the rate of oxidation of SO_2 decreases as the pH falls below the value of pK_1 for reaction (5.4). As the oxidation of SO_2 results in the formation of H_2SO_4, which is a stronger acid than H_2SO_3, the pH of the reaction solution decreases as oxidation proceeds. The rate of oxidation thus decreases with time. In order to provide a mechanism to explain the relatively high rate of SO_2 oxidation in atmospheric water some workers have suggested that the buffer capacity of dissolved CO_2 and NH_3 may be sufficient to maintain solution pH at a level where oxidation is quite rapid. The effect of ammonia in solution was found to be quite dramatic—with an SO_2 concentration of $20\ \mu g\ m^{-3}$ and no NH_3 a sulphate concentration of $0.5\ \mu g\ m^{-3}$ was calculated for 24 hours oxidation, while in the presence of $5\ \mu g\ m^{-3}$ NH_3 the sulphate concentration was calculated as about $60\ \mu g\ m^{-3}$ for the same time. The concentration parameters for both SO_2 and NH_3 are reasonable for atmospheres found in populated areas.

Ammonium sulphate, the product of SO_2 oxidation in the presence of NH_3, has been identified as a major constituent of the atmospheric aerosol. Because of its deliquescent behaviour it is of considerable importance in the formation of hazes and mists which have a marked influence on visual range.

Photo-oxidation

Here we are to consider reactions of SO_2 that, in the gas phase and in the presence of solar radiation, produce SO_3 and hence H_2SO_4. The photochemical dissociation of SO_2 (reaction 5.18) is not possible under atmospheric conditions as it has an energy requirement of $565\ kJ\ mol^{-1}$ which cannot be satisfied by the absorption of solar radiation.

$$SO_2 + h\nu \rightarrow SO + O \qquad (5.18)$$

An alternative mode of photo-oxidation of SO_2 is through an electronically-excited state of SO_2 produced when SO_2 absorbs solar radiation. The absorption spectrum of SO_2 shows two bands in the wavelengths of tropospheric solar radiation. A weak absorption with a maximum at 388 nm produces a triplet state ($^3SO_2$) while a strong absorption with a maximum at 294 nm produces a singlet state ($^1SO_2$). The reactions of these electronically-excited states of SO_2 in pure SO_2 are summarized in reactions (5.19)–(5.27).

$$SO_2 + h\nu \rightarrow {}^1SO_2 \qquad (294\ \text{nm}) \qquad (5.19)$$

$$^1SO_2 + SO_2 \rightarrow (2SO_2) \qquad (5.20)$$

$$^1SO_2 + SO_2 \rightarrow {}^3SO_2 + SO_2 \qquad (5.21)$$

$$^1SO_2 \rightarrow SO_2 + h\nu_f \qquad (5.22)$$

$$^1SO_2 \rightarrow SO_2 \qquad (5.23)$$

$$^1SO_2 \rightarrow {}^3SO_2 \qquad (5.24)$$

$$^3SO_2 \rightarrow SO_2 + h\nu_p \qquad (5.25)$$

$$^3SO_2 \rightarrow SO_2 \tag{5.26}$$

$$^3SO_2 + SO_2 \rightarrow (2SO_2) \tag{5.27}$$

Reactions (5.21) and (5.24) are intercrossing reactions by which $^3SO_2$ may be formed from $^1SO_2$. They are very important as it has been shown that $^3SO_2$ is the reactive species in SO_2 photo-oxidation and, as SO_2 has only a weak absorption band for the direct formation of $^3SO_2$, they provide a pathway for the formation of significant amounts of $^3SO_2$.

Reactions (5.22), (5.23), (5.25), and (5.26) represent the decay to ground-state SO_2 from both excited states either by the emission of radiation or by radiationless decay. The product, $(2SO_2)$, of reactions (5.20) and (5.27) can be any chemical product that does not regenerate an excited state of SO_2 and may be SO_3 or a compound that produces SO_3 e.g. the very reactive tetroxide, SO_4.

In the atmosphere SO_2 is only a trace constituent hence other interactions similar to (5.20), (5.21), and (5.27) must also be considered:

$$^1SO_2 + M \rightarrow SO_2 + M \tag{5.28}$$

$$^1SO_2 + M \rightarrow {}^3SO_2 + M \tag{5.29}$$

$$^3SO_2 + M \rightarrow products \tag{5.30}$$

$$^3SO_2 + M \rightarrow SO_2 + M \tag{5.31}$$

where M is some molecule other than SO_2, usually N_2, O_2, or H_2O because of their high abundance, or olefins because of their very high reactivity. Reaction (5.30) is the most important reaction for the production of SO_3. The exact nature of this reaction is not clear, however the following postulated reactions have considerable support from workers in the field:

$$^3SO_2 + O_2 \rightarrow SO_3 + O \tag{5.32}$$

$$O + O_2 \rightarrow O_3 \tag{5.33}$$

Ozone has never been found in SO_2 photo-oxidation, perhaps because of the analytical difficulty of its determination in the presence of SO_2.

Oxidation on aerosols

Although SO_2 has been shown to be rapidly sorbed by aerosols of Fe_3O_4, Al_2O_3, and CaO the subsequent reactions of the gas are not known in any detail. Sulphate has been found in the solid particulate material in the atmosphere but it has not been shown that it was formed *in situ* on the aerosol by oxidation of SO_2. It could well have arisen from coagulation of a H_2SO_4 aerosol droplet with a solid particulate aerosol.

Comparison of the oxidation mechanisms

When making a comparison of the mechanisms for SO_2 oxidation in the atmosphere a wider range of parameters must be considered than were considered in laboratory studies. For instance, the photochemical mechanism will obviously have a zero oxidation rate at night and the rate will also vary with the intensity of solar radiation during the day. Taking these factors into consideration an SO_2 oxidation rate of about 0·15 per cent ks^{-1} integrated over a twenty-four hour period can be deduced for photo-oxidation.

In the case of SO_2 oxidation in droplets in the atmosphere consideration must be given to the rate of dissolution of SO_2 into the droplets, the rate of dissolution of metal oxides to produce metal ions for catalysis, and to the relative humidity of the atmosphere which controls droplet formation. No estimates have been made taking into account the rate of production of metal ions for catalysis. All calculations have been based on the assumption that all of the metal under consideration is available in soluble form to assist SO_2 oxidation. The maximum oxidation rates calculated on this basis occur under fog conditions and may reach a rate of 0·6 per cent ks^{-1} with Mn^{II} as a catalyst. It is possible to calculate rates of up to 25 per cent ks^{-1} with Fe^{III} as a catalyst in a chimney plume. The oxidation rates reported in Table 5.1 are thus adequately explained by this mechanism.

The mean residence time of SO_2 in the atmosphere has been derived from data collected at monitoring stations in various parts of the world. The results vary from twelve hours to six days. It can thus be seen that photochemical oxidation and solution oxidation separately, or together, could account for the longer observed residence times. The shorter residence times must be influenced by factors other than SO_2 oxidation e.g. absorption of SO_2 into natural waters, deposition of SO_2 on to vegetation and soils, etc.

Deposition of SO_2 into natural waters

In Chapter 3 the basic physical concepts for gaseous exchange between the gas phase and the aqueous phase were discussed with reference to CO_2. In that discussion the influence of the gas-phase boundary layer was considered to be insignificant. Experiments with SO_2 have shown that the exchange coefficient is constant above pH 6 while it decreases markedly with pH below pH 6. This implies that the exchange coefficient is influenced by the change of ionic species in solution and it can be shown that, at most pH values, diffusion through the gas-phase boundary layer must be considered.

The resistance R to SO_2 exchange from the gas phase to solution is, by analogy with Ohm's Law, the sum of the resistance to exchange through the gas-phase boundary layer r_g and the resistance to exchange through the solution phase boundary layer r_1, i.e.

$$R = r_g + r_1. \tag{5.34}$$

Continuing the analogy with electricity we find that the exchange coefficient is analogous with conductance hence eqn (5.34) may be written:

$$\frac{1}{k} = \frac{1}{k_1} + \frac{1}{Hk_g} \qquad (5.35)$$

where k is the overall exchange coefficient, k_1 is the exchange coefficient for transport through the solution phase, and k_g is the exchange coefficient for transport through the gas phase. H, Henry's Law constant, is applied to bring the exchange coefficients to the same concentration basis.

The value for k_g depends solely upon windspeed and may thus vary from 0.3 cm s^{-1} in calm conditions up to 1.5 cm s^{-1} in turbulent conditions. On the other hand, k_1 is greatly influenced by the ionic species in solution which vary markedly with pH. At pHs where ionic species predominate the enhancement α, (see Chapter 3), to the exchange coefficient for aquated molecular SO_2 alone is very large indeed, e.g. at pH 2 where ionic S^{IV} species are in low concentration α is about 3 whereas at pH 6 α is about 3000. The overall effect of this can be seen in Table 5.3. At pH 2·8, under the wind

TABLE 5.3

Calculated resistance to SO_2 *exchange into aqueous solutions of varying* pH
(after Liss 1971)

Solution pH	Ionic enhancement factor, α	Liquid-layer resistance, r_1 (cm s^{-1})	Gas-layer resistance, r_g (cm s^{-1})	Total resistance, R (cm s^{-1})
2	2·7	1·00	0·25	1·25
3	18	0·15	0·25	0·40
4	169	0·02	0·25	0·27
5	1376	0·002	0·25	0·25
6	2884	0·001	0·25	0·25
7	2966	0·001	0·25	0·25
8	2967	0·001	0·25	0·25
9	2967	0·001	0·25	0·25

conditions chosen for the calculation, $r_1 = r_g$. Below pH 2·8 r_1 is dominant, and above r_g is dominant. Most natural waters are of pH 4–9 hence it can be concluded that at almost all environmental air–water interfaces, SO_2 exchange rate is controlled by diffusion through the gas-phase boundary layer. Water vapour exhibits similar behaviour, while O_2 and CO_2 exchange with natural waters are controlled by diffusion through the liquid phase boundary layer.

As SO_2 has a high solubility (at 15°C 45 vols SO_2 dissolve in 1 vol H_2O) virtually all SO_2 reaching the surface of natural waters will dissolve. Sea water has a pH of about 8 and is buffered by the carbonate–bicarbonate system. It can thus be concluded that the oceans are an important sink for SO_2. Recently a direct measurement of the rate of absorption of SO_2 by seawater has been made. Under the most turbulent conditions the value for the total resistance to exchange was 0.7 s cm^{-1}. This value may be used in the following manner to estimate the total uptake of SO_2 from the atmosphere by the oceans.

The total resistance R is related to the rate of absorption by:

$$R = \frac{1}{V_g} \tag{5.36}$$

where V_g, the velocity of deposition, is defined as

$$V_g = \frac{M}{[SO_2]_g \times t} \tag{5.37}$$

where M = Mass of SO_2 (μg) sorbed per unit area (cm^2)
$[SO_2]_g$ = Gas-phase concentration of SO_2 ($\mu g \text{ cm}^{-3}$)
t = Time of gas exposure to solution (s).

Data necessary for calculation:

Number of seconds in one year = 3.156×10^7 s

Average concentration of S as SO_2 over oceans = $2.5 \times 10^{-7} \mu g \text{ cm}^{-3}$

$V_g = \dfrac{1}{R} = 1.4 \text{ cm s}^{-1}$

Area of oceans = $3.7 \times 10^{18} \text{ cm}^2$

Mass of S sorbed as SO_2 = $1.4 \times 2.5 \times 10^{-7} \times 3.156 \times 10^7 \times 3.7 \times 10^{18} \mu g$
$= 41 \times 10^6$ tonnes.

This represents a greater absorption of SO_2 by the oceans than is given for the sulphur cycle (Fig. 5.1). The difference arises only from the value of V_g used. The sulphur cycle figure was obtained using an estimated V_g of 0.9 cm s^{-1} while the above calculation used a measured V_g of 1.4 cm s^{-1}. Greater SO_2 absorption by the ocean suggests a lower emission of H_2S or sea-spray sulphate necessary to balance the cycle. The former is more attractive as it has often been considered that H_2S would be rapidly oxidized in sea water and thus be unavailable for exchange to the atmosphere.

Deposition of SO_2 to vegetation and soil

Although most plants are damaged by SO_2 when exposed to concentrations of about 1 mg kg^{-1} for several hours, almost all of the world's vegetation receives non-damaging concentrations of SO_2. Plant uptake of SO_2 is almost completely through the leaf pores (stomata) which control gas exchange between the interior of the leaf and the atmosphere. The stomata are under the physiological control of the plant and open or close depending on the plant's need for CO_2 or water exchange. When the atmosphere is at a low relative humidity the stomata close to conserve water while high relative humidity and high sunlight intensity cause stomatal opening so that CO_2 exchange and photosynthesis are promoted. Maximum SO_2 uptake is found at 100 per cent relative humidity in the atmosphere and corresponds to maximum stomatal opening.

As was the case with SO_2–sea water exchange it is possible to express SO_2 uptake by leaves as a resistance to total gaseous exchange, R.

$$R = r_g + r_s + r_{mes} \qquad (5.38)$$

R and r_g are as defined in (5.34), r_s is the resistance to exchange through the stomata, and r_{mes} is the resistance to exchange at the damp cell (mesophyll) surfaces within the leaves.

The minimum value for R has been found to be 2.8 s cm^{-1}. Values for r_g and r_s can be calculated for a given set of experimental conditions. The sum of these two resistances under conditions of maximum SO_2 exchange approaches the value of R. This implies that the resistance to SO_2 exchange at the damp mesophyll cell surfaces is very low and, by analogy with r_l (eqn 5.34), is not controlled by chemical reaction within the mesophyll cells. The reverse is true for CO_2, where it is found that r_{mes} is an important contributor to R.

The reciprocal of R gives the velocity of deposition (V_g) for maximum SO_2 exchange to plant leaves. This has a value of 0.36 cm s^{-1}. In order to calculate the total deposition of SO_2 to plant leaves per unit area of land it is necessary to estimate the total leaf area per unit area of land. This varies with plant species but a factor of four times the leaf area (top plus underside) per unit area of land appears to be the average. The overall value of V_g for SO_2 deposition to plant leaves, per unit area of land, is thus about 1.5 cm s^{-1}. A calculation similar to that carried out for SO_2 absorption by sea water leads to a value for annual SO_2 intake by vegetation similar to that given in the Sulphur Cycle (Fig. 5.1).

It is also possible to calculate that vegetation is able to be supplied annually with 50 mg m^{-2} of S for each $1 \text{ } \mu\text{g m}^{-3}$ of SO_2 in the atmosphere. This would be marginally sufficient as the sole source of sulphur for a wide range of crops. Many soils in the world are deficient in sulphur hence the transfer of sulphur in fuels directly to plants in the form of atmospheric SO_2 could be of some importance to world agriculture especially in regions where the application of chemical fertilizers is not carried out for economic reasons.

The direct absorption of SO_2 by soils is not included in the Sulphur Cycle. It is known that soils absorb SO_2 and that the rate depends very much upon the moisture content of the soil. At high moisture contents the SO_2 absorption rate is maximal. Micro-organisms also have an effect, causing an increase in absorption rate as their population increases. In calculating the mass of SO_2 absorbed by vegetation it was assumed that the whole of the surface of the earth was covered by plants. The uptake of SO_2 by soils is probably accounted for in the total sulphur cycle by this simplification.

Damaging effects of SO_2

Effects on man

Few data are available on the effects of SO_2 air pollution on human health. The classic epidemiological study of the London smog of 5–8 December 1952 showed an excess of 3500–4000 deaths above the predicted value. In this smog high concentrations of particulate matter as well as SO_2 were measured. It is generally accepted that the combined effect of both constituents on the respiratory tract was responsible for the excess deaths.

Laboratory data indicate that SO_2 has the potential of slowing down ciliary movements in the respiratory tract. The cilia act to clear micro-organisms and toxic particles from the respiratory tract. If these irritants reach the lungs they may cause acute respiratory problems. There is no evidence that SO_2 at air pollutant concentrations in the absence of high concentrations of particulate material can cause adverse effects to humans in a normal state of health.

Effects on plants

It is often found with gaseous air pollutants that plants are damaged at much lower concentrations than those at which human health is affected. Sensitivity to SO_2 varies with plant species e.g. alfalfa, barley, cotton and wheat are listed as sensitive while potato, onion, corn and maple are listed as resistant. It appears that even the most sensitive species shows no visible response to an exposure of less than $100 \ \mu g \ m^{-3}$ of SO_2, for an indefinite period. Continuous chronic exposure to SO_2 concentrations of as low as $300 \ \mu g \ m^{-3}$ lead to leaf damage in some sensitive species, while short-time acute exposures may cause damage to leaves at as low as $700 \ \mu g \ m^{-3} \ SO_2$.

Sufficient knowledge has now accumulated to enable the identification of SO_2 damage by the physiological state of the leaves e.g. discolourations, necroses etc. Little is known of the biochemical basis for the damage although recently an aldehyde–hydrogensulphite adduct has been isolated from rice plants treated with SO_2.

$$\begin{array}{c} COO^- \\ | \\ HC{=}O \end{array} \ \text{glyoxalate} \ \xrightarrow{\ HSO_3^-\ } \ \begin{array}{c} COO^- \\ | \\ HC{\cdot}SO_3^- \\ | \\ OH \end{array} \ \begin{array}{l} \text{glyoxalate} \\ \\ \text{hydrogensulphite} \end{array} \quad (5.39)$$

Glyoxalate hydrogensulphite is a competitive inhibitor of the enzyme glycolic oxidase which catalyzes the oxidation of glycolate:

$$\begin{array}{c} COO^- \\ | \\ CH_2OH \end{array} + O_2 \ \xrightarrow{\ \text{glycolic oxidase}\ } \ \begin{array}{c} COO^- \\ | \\ CHO \end{array} + H_2O_2 \quad (5.40)$$

glycolate   glyoxalate

Reaction (5.40) is part of the glycolic pathway which is associated with the process of photorespiration in plants. Independent evidence indicates that glycolic oxidase inhibition occurs in barley plants treated with SO_2. The importance of these observations awaits further experimental elucidation.

Effects on building materials

Limestone and marble have been used as building materials for many centuries. The damage to limestone in particular is obvious in industrial cities where severe erosion of limestone building blocks and ornamentation may be seen.

The mass of SO_2 absorbed by limestone is increased as relative humidity increases. The absorbed gas is oxidized to sulphate and becomes part of the $CaCO_3$ matrix. The molecular volume of $CaSO_4$ is greater than that of $CaCO_3$ hence mechanical stresses on the molecular scale arise. The accumulated effect of these stresses is to cause flaking-off of the limestone. Further, $CaSO_4$ has a higher solubility in rain water than $CaCO_3$ (209 mg ml^{-1} and 1·4 mg ml^{-1} respectively) and is thus readily leached out.

Works of art, especially frescoes, are susceptible to SO_2 damage by the same mechanism. The true frescoe is a pigmented lime plaster which may also be converted to $CaSO_4$ in the presence of SO_2. Marble is less susceptible to SO_2 attack probably because of its low porosity.

Unpainted timber absorbs SO_2 and this may cause some degradation due to the reaction of sulphite with lignins *cf.* the sulphite-pulping process for the production of cellulose from wood. Damage by other weathering processes probably outweighs that due to SO_2. Painted timbers may also be attacked by SO_2 as it has been found that many paint films are permeable to SO_2.

Effects on paper

Of the many objects that are stored by man probably the most important are books. The yellowing and loss of mechanical strength of paper is accelerated in books stored in industrial cities. SO_2 has been clearly shown to be the major contributor to the degradation of paper stored in these places.

The usual phenomenon of increased SO_2 sorption with increased atmospheric humidity is also found with paper. Studies with radioactive SO_2 have shown that high concentrations of SO_2 are found about metallic impurities left in the paper from the manufacturing process. It is supposed that SO_2 is rapidly oxidized to H_2SO_4 at these points and that the basic degradation process is one of acid hydrolysis of cellulose. In addition, it is thought that the formation of lignosulphonic acids by the reaction between SO_2 in the surface moisture on the paper and lignins within it, also has an effect.

The handling of paper results in sweat deposits being left on the paper. Such deposits show a great capacity to sorb SO_2 hence damage to the outer edges of books may not only be due to abrasive actions but also to high

H_2SO_4 concentrations within sweat deposits. Similar effects occur on wallpapers but wallpapers are usually changed for social reasons before significant damage can be seen.

Effects on leather

Damage to leather by SO_2 has been known since at least 1843 when Faraday suggested that the deterioration of the leather armchairs of his London club was due to SO_2. Leather bookbindings also suffer SO_2 damage which takes the form of cracking and loss of flexibility and mechanical strength.

It has been shown that SO_2 uptake by leather is almost completely controlled by the rate of SO_2 diffusion to the surface. At the surface oxidation results in the formation of H_2SO_4 which may cause acid hydrolysis of leather protein. The amount of SO_2 sorbed by leather can be greatly reduced by providing a surface coating of nitrocellulose or polyurethane.

Effect on metals

Observations on the rate of corrosion of test panels of many metals exposed to the atmosphere at a variety of sites have shown that, with most metals, greatest corrosion occurs in industrial atmospheres. In these conditions the rate of corrosion correlates very closely with the SO_2 concentration in the atmosphere and with the time of wetness of the metal surfaces.

The atmospheric corrosion of iron and steel may be explained by an electrochemical mechanism. In the atmosphere iron and steel are always covered with a thin layer of Fe_3O_4 which itself is covered by a film of its oxidation product FeOOH (Fig. 5.2). At the surface of the pure metal the anodic reaction (5.41) occurs, while at the Fe_3O_4/FeOOH interface cathodic reduction followed by atmospheric oxidation occurs (reactions 5.42 and 5.43).

$$Fe \rightarrow Fe^{2+} + 2e \qquad (5.41)$$

$$Fe^{2+} + 8FeOOH + 2e \rightarrow 3Fe_3O_4 + 4H_2O \qquad (5.42)$$

$$3Fe_3O_4 + 0.75O_2 + 4.5H_2O \rightarrow 9FeOOH \qquad (5.43)$$

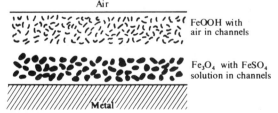

Air

FeOOH with
air in channels

Fe_3O_4 with $FeSO_4$
solution in channels

Metal

FIG. 5.2. Schematic representation of the electrochemical mechanism of atmospheric rusting of iron in the presence of SO_2 (after Evans 1972).

The overall effect is to increase the amount of rust, FeOOH, by one eighth by transferring iron from the metal to the surface rust layer.

As the anodic and cathodic sites for the corrosion process are separate in space, conductors are needed to complete the electrochemical circuit. It is supposed that magnetite, Fe_3O_4, is the electronic conductor and that $FeSO_4$ which exists in solution in the Fe_3O_4 layer acts as the ionic conductor. The role of SO_2 in the corrosion process is thus to provide SO_4^{2-} by its oxidation after sorption on the metal. The time of wetness relationship is explained by the necessity for $FeSO_4$ to remain in solution in the Fe_3O_4 layer so that the electrochemical circuit is completed. When the layer dries out the circuit is disrupted and corrosion ceases.

Aluminium is susceptible to attack in both industrial and marine atmospheres. The attack in industrial atmospheres is thought to be due to the formation of H_2SO_4 from SO_2 absorbed on the surface. The acid breaks down the oxide film that normally provides a protection against atmospheric attack, and corrosion proceeds. A similar process occurs on the surface of zinc where the protective film of basic zinc carbonate is dissolved by H_2SO_4 formed from SO_2 absorbed on the zinc surface.

6. Nitrogen oxides and photochemical smog

NITROGEN oxides play a very important role in the formation of the type of smog that was first recognized in the Los Angeles district, and which is now referred to as photochemical smog. This type of smog is typified by photochemical reactions brought about by the action of solar radiation on motor vehicle exhausts. The most damaging constituents of the smog are nitric oxide, nitrogen dioxide, ozone, and peroxoacyl nitrates. Vegetation is damaged by very low concentrations of some of these constituents while slightly higher concentrations cause unpleasant effects to humans, especially to the respiratory system.

Nitrogen forms oxides corresponding to each of its known oxidation states (Table 6.1). Of these oxides, only N_2O, NO, and NO_2 appear at measurable concentrations in the unpolluted atmosphere. The equilibria involving N_2O_3, N_2O_4, and N_2O_5 are all heavily in favour of dissociation at atmospheric temperatures and partial pressures.

TABLE 6.1
Oxides of nitrogen

Oxide	Formula	Stability in the atmosphere
Dinitrogen oxide	N_2O	Stable gas
Nitrogen oxide	NO	Stable gas
Dinitrogen trioxide	N_2O_3	Unstable gas
		$N_2O_3 \rightleftharpoons NO + NO_2$
Nitrogen dioxide	NO_2	Stable gas
Dinitrogen tetroxide	N_2O_4	Unstable gas
		$N_2O_4 \rightleftharpoons 2NO_2$
Dinitrogen pentoxide	N_2O_5	Unstable gas
		$N_2O_5 \rightleftharpoons N_2O_3 + O_2$
Nitrogen trioxide	NO_3	Unstable gas (never isolated)

Dinitrogen oxide (nitrous oxide)

This gas was first identified in the atmosphere by its infrared absorption spectrum as recently as 1939. The mean atmospheric concentration deduced from the infrared spectral data is 0.25 mg kg^{-1}. The concentration of N_2O remains reasonably constant up to the tropopause and then decreases with altitude because of photodissociation reactions (6.1) and (6.2).

$$N_2O + hv \longrightarrow N_2 + O \quad (\lambda < 337 \text{ nm}) \qquad (6.1)$$
$$N_2O + hv \longrightarrow NO + N \quad (\lambda < 250 \text{ nm}) \qquad (6.2)$$

Photodissociation is not found to any extent in the troposphere as the solar radiation spectrum in that layer has a sharp cut-off in the ultraviolet region at 390 nm due to absorption in the ozone layer (see Fig. 3.1).

The principal source of atmospheric N_2O is the soil. Micro-organisms in the soil are able to degrade protein nitrogen to nitrogen gas and N_2O. The maximum rate of production of N_2O thus depends upon optimum conditions for the activity of soil micro-organisms, and a plentiful supply of protein. These conditions vary from season to season, hence it is not surprising that a seasonal variation in the concentration of N_2O in the atmosphere is found.

Micro-organisms are also capable of reducing N_2O under anaerobic conditions. It is thought that N_2O that diffuses to the deep ocean is destroyed in this manner. The most important processes for the destruction are, however, the stratospheric reactions (6.1) and (6.2). The normal cycle of atmospheric N_2O is thus production in the soil followed by diffusion to the stratosphere where photodissociation occurs. The calculated mean residence time for N_2O is subject to some conjecture. Values between 4 years and 70 years have been obtained.

N_2O is not recognized as an air pollutant. Its importance in the field of air pollution chemistry is in its photodissociation to NO which is an important pollutant gas.

Nitrogen oxide (nitric oxide) and nitrogen dioxide

These two oxides will be discussed together because of their relationship through equilibrium (6.3)

$$2NO + O_2 \underset{k_{-1}}{\overset{k_1}{\rightleftharpoons}} 2NO_2 \rightleftharpoons N_2O_4 \qquad (6.3)$$

In the solid state nitrogen dioxide exists entirely as colourless N_2O_4 while in the vapour state at 100°C the composition at equilibrium is 90 per cent NO_2 and 10 per cent N_2O_4. For the purposes of further discussion nitrogen dioxide will be regarded as being the brown gaseous NO_2.

The production of NO_2 from NO in the atmosphere takes place rather slowly as reaction (6.3) is second order in NO concentration. At an NO concentration of 0.1 mg kg^{-1} the half life for the reaction is about 4 Ms, which is in marked contrast to the almost immediate production of brown NO_2 fumes from high concentrations of NO. The NO_2 produced absorbs strongly in the ultraviolet region, dissociating to NO and atomic oxygen (reaction 6.4). Reaction (6.5) accounts for most of the atomic oxygen produced. The overall effect is thus to produce equal concentrations of NO and O_3 from the NO_2 present in the atmosphere. These compounds react together (reaction 6.6) thus completing a cyclic reaction,

$$NO_2 + hv \longrightarrow NO + O \qquad (\lambda < 380 \, nm) \qquad (6.4)$$

$$O + O_2 + M \longrightarrow O_3 + M \qquad (6.5)$$

(M is any third body, usually N_2 or O_2)

$$NO + O_3 \rightarrow NO_2 + O_2 \tag{6.6}$$

This produces a situation where the concentrations of NO and NO_2 remain constant being controlled by the probability of reaction (6.6). The addition of any other compound that reacts with atomic oxygen in particular will disrupt this pseudo-equilibrium situation. Such is the case in photochemical smog.

The predominant sources of NO are oxidation of NH_3 and high-temperature combustion processes. Both of these are tropospheric sources. Minor sources are found in the stratosphere and thermosphere. In the thermosphere NO is formed by the reaction of oxygen with atomic nitrogen (reaction 6.8)

$$N_2 + hv \rightarrow N + N \tag{6.7}$$

$$N + O_2 \rightarrow NO + O \tag{6.8}$$

This is thought to be the principal reaction for the removal of atomic nitrogen from the atmosphere. Reaction (6.2) is also of some importance in the production of stratospheric NO.

Both NO and NO_2 are found in combustion gases, with NO predominating as its formation is favoured at high temperatures. NO is formed in post-flame combustion gases from nitrogen and oxygen in the air used to combust the fuel (reactions 6.9–6.12).

$$CO + OH \rightleftharpoons CO_2 + H \tag{6.9}$$

$$H + O_2 \rightleftharpoons OH + O \tag{6.10}$$

$$O + N_2 \overset{k_1}{\rightleftharpoons} NO + N \tag{6.11}$$

$$O_2 + N \rightleftharpoons NO + O \tag{6.12}$$

Reactions (6.9) and (6.10) occur together in the combustion gases and are largely responsible for the concentration of atomic oxygen. If a steady concentration of nitrogen atoms and combustion under excess air conditions are assumed, it may be shown that the rate of production of NO approximates to eqn (6.13).

$$\frac{d[NO]}{dt} = 2k_1[O][N_2] \tag{6.13}$$

The value of k_1 in reaction (6.11) is $1 \times 10^{11} \exp(-75\,400/RT)\,\mathrm{dm^3\,mol^{-1}\,s^{-1}}$. Application of this to eqn (6.13) shows that the temperature is the most significant factor in the production of NO under normal combustion conditions. The higher the temperature, the greater the production of NO. The production of NO per unit mass of fuel combusted follows the order coal > oil > natural gas, as this is the order of average combustion temperature. In terms of mass emitted, motor vehicles are the most important source of NO as the internal combustion engine operates at a high temperature.

Photochemical smog

The most important atmospheric reactions involving NO and NO_2 occur in the group of reactions known as photochemical smog. A typical photochemical smog occurs in warm sunny weather and it is characterized by a haze, ozone formation, eye irritation, and damage to vegetation. This smog was first recognized in the Los Angeles district hence the following general outline of photochemical smog will refer to the Los Angeles situation.

The geography of the Los Angeles basin is such that the district may be likened to a giant chemical reaction flask. The walls of the 'flask' are made up of the floor of the valley and mountains which occupy three sides of the valley. The fourth side of the valley is open to the ocean but the prevailing wind from the sea effectively closes this wall of the 'flask'. A 'lid' is provided by frequent temperature inversions at 200–500 m. A temperature inversion is found when a layer of warm air overlies cool air adjacent to the ground. This prevents turbulent convective mixing of the air adjacent to the ground with the remainder of the troposphere and thus all gases emitted into this air mass are trapped. A large-scale temperature inversion occurs at the tropopause which effectively separates the troposphere from the stratosphere (Fig. 1.1).

To this reaction 'flask' are added the photochemical smog reagents—exhaust gases from the operation of a large number of motor vehicles. The energy required for the photochemical smog reaction to proceed comes from the solar radiation spectrum. Los Angeles is well-known for its high number of sunlight hours.

The major changes that may be measured in the atmosphere during a photochemical smog episode are illustrated in Fig. 6.1. The sequence of events commences with the injection of hydrocarbons and NO into the atmosphere from the exhaust systems of motor vehicles. With increasing sunlight intensity the concentration of NO decreases, while the concentrations of NO_2 and the aldehydes increase. A decrease in the NO_2 concentration accompanies the appearance of significant O_3 levels which, after midday, show a decrease as do the levels of hydrocarbons and aldehydes. There is no observed increase in NO or hydrocarbon concentrations during the evening traffic peak.

The key reactant in the formation of photochemical smog is NO_2. In conditions where hydrocarbons are absent we have seen that photodissociation sets up a pseudo-equilibrium concentration of NO and NO_2 (reactions 6.4–6.6). In the presence of hydrocarbons this equilibrium is disrupted by a chain reaction initiated by the reaction of atomic oxygen and O_3 with hydrocarbons.

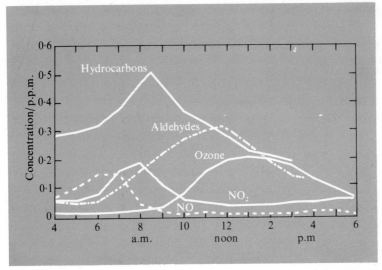

FIG 6.1. Average concentrations of some atmospheric constituents during days of eye irritation in Los Angeles.

The polluted urban atmosphere contains about one hundred different hydrocarbons, the most reactive of which are the olefins. The result of atomic oxygen attack on olefins is the production of two free radicals. In the case of propylene, the first step in the reaction is understood to be the addition of O to the double bond to produce an activated complex (6.14).

$$
\begin{array}{c}
H \\
{}^{\diagdown}C=C{}^{\diagup} \quad + O \quad \rightarrow \quad \left[\begin{array}{c} H \\ {}^{\diagdown}C-C{}^{\diagup} \\ H_3C \end{array} \begin{array}{c} H \\ O \\ H \end{array} \right]
\end{array}
\tag{6.14}
$$

The activated complex may fragment in two different ways (reactions 6.15 and 6.16).

$$
\left[\begin{array}{c} H \\ C-C-O \\ H_3C \end{array} \begin{array}{c} H \\ \\ H \end{array} \right] \rightarrow H_3C-\overset{H}{\underset{H}{C}}\cdot + \cdot C\overset{O}{\underset{H}{\diagup}}
\tag{6.15}
$$

$$
\left[\begin{array}{c} H \\ C-C-O \\ H_3C \end{array} \begin{array}{c} H \\ \\ H \end{array} \right] \rightarrow H_3C\cdot + H_3C-C\overset{O}{\diagup}\cdot
\tag{6.16}
$$

Reaction (6.15) is the more likely as it involves less rearrangement of the activated complex than does reaction (6.16). CHO· rapidly forms formaldehyde and $CH_3CO·$ rapidly forms acetaldehyde. Reactions (6.15) and (6.16) are the initiation reactions of a chain reaction (reactions 6.17–6.21).

$$CH_3· + O_2 \rightarrow CH_3O_2· \tag{6.17}$$

$$CH_3O_2· + NO \rightarrow CH_3O· + NO_2 \tag{6.18}$$

$$CH_3O· + O_2 \rightarrow HCHO + NO_2· \tag{6.19}$$

$$HO_2· + NO \rightarrow OH· + NO_2 \tag{6.20}$$

$$C_3H_6 + OH· \rightarrow CH_3CH_2· + H_2O \tag{6.21}$$

It must be emphasized that the chain propagation reactions (6.17) to (6.21) are subject to some conjecture; however they do explain the formation of many of the compounds found in the laboratory studies of photochemical smog. The chain reaction enables a rapid oxidation of NO to NO_2 by alkoxyl (RO·) and peroxoalkyl ($RO_2·$) radicals, thus atomic oxygen and thus O_3 are conserved. This offers some explanation for the observed changes in photochemical pollutant gas concentration throughout the day (Fig. 6.1). Other alkenes can also follow this reaction sequence, the overall rate of their oxidation varying with their structure.

The initial 'equilibrium' concentrations of NO and NO_2 are controlled by the photolytic NO_2 cycle (reactions 6.4–6.6). As the atmospheric hydrocarbon concentration increases, due to motor vehicle operation, the photolytic NO_2 cycle is disrupted and NO is oxidized to NO_2 by the hydrocarbon chain reaction (eqns 6.15–6.21). The low steady-state O_3 concentration found in the photolytic NO_2 cycle thus increases because O_3 is not being used to oxidize NO to NO_2. The hydrocarbon concentration decreases because of its participation in NO oxidation, and the aldehyde concentration increases as it is a product of the hydrocarbon chain oxidation of NO. Of course, the NO concentration decreases and the NO_2 concentration increases as a result of these reactions.

It is now necessary to consider other interactions of the photochemical smog gases in order to explain the further changes in the gas concentrations shown in Fig. 6.1. Hydrocarbon radicals may react with NO_2 to form organic nitrates (reaction 6.22) or peroxoacyl nitrates (reactions 6.23–6.25).

$$RO· + NO_2 + M \rightarrow RNO_3 + M \tag{6.22}$$

$$O + R'H \rightarrow R''· + RCO· \tag{6.23}$$

$$RCO· + O_2 \rightarrow R(CO)O_2· \tag{6.24}$$

$$R(CO)O_2· + NO_2 \rightarrow R(CO)O_2NO_2 \tag{6.25}$$

(M is any third body, usually N_2 or O_2; R, R', and R'' are any hydrocarbon chains.)

Reaction (6.22) is a termination reaction for the hydrocarbon chain oxidation of NO to NO_2. Reaction (6.23) is the first portion of a reaction represented by eqns (6.15) and (6.16). The peroxoacyl radical $R(CO)O_2$ formed by the oxidation of the product of reaction (6.23) is basic to the production of peroxoacyl nitrates (PANs). The PANs are very important in photochemical smog because of their considerable biological reactivity. They cause plant leaf damage at very low concentrations and cause eye and respiratory irritation in humans at concentrations as low as 0.5 mg kg^{-1}. The best-known PAN is peroxoacetylnitrate ($CH_3C \cdot OO \cdot NO_2$).

The reactions represented by eqns (6.22)–(6.25) thus give some explanation for a decrease in the NO_2 concentration and a further decrease in the hydrocarbon concentration. The formation of the PANs is also explained.

It remains now to consider the reactions between O_3 and the olefinic hydrocarbons. These reactions result initially in the formation of an ozonide intermediate (reaction 6.26).

$$\begin{array}{c}R \\ \diagdown \\ \diagup \\ H\end{array} C{=}C \begin{array}{c} R' \\ \diagdown \\ \diagup \\ H \end{array} + O_3 \rightarrow \left[\begin{array}{c} R \\ \diagdown \\ \diagup \\ H \end{array} C \begin{array}{c} O \\ \diagdown \\ O{-}O \end{array} C \begin{array}{c} R' \\ \diagdown \\ \diagup \\ H \end{array} \right] \qquad (6.26)$$

where R and R′ are hydrocarbon chains.

The ozonide intermediate may cleave in one of two ways (reactions 6.27 and 6.28)

$$\left[\begin{array}{c} R \\ \diagdown \\ \diagup \\ H \end{array} C \begin{array}{c} O \\ \diagdown \\ O{-}O \end{array} C \begin{array}{c} R' \\ \diagdown \\ \diagup \\ H \end{array} \right] \rightarrow R\dot{C}HO \cdot + R'\dot{C}HOO \cdot \qquad (6.27)$$

$$\left[\begin{array}{c} R \\ \diagdown \\ \diagup \\ H \end{array} C \begin{array}{c} O \\ \diagdown \\ O{-}O \end{array} C \begin{array}{c} R' \\ \diagdown \\ \diagup \\ H \end{array} \right] \rightarrow R\dot{C}HOO \cdot + R'\dot{C}HO \cdot \qquad (6.28)$$

The biradical products from reactions (6.27) and (6.28) may be written as zwitterions which can undergo a variety of decomposition reactions (such as reactions 6.29–6.31).

$$R'\overset{+}{C}HOO^- \rightarrow R'O \cdot + CHO \cdot \qquad (6.29)$$

$$R'\overset{+}{C}HOO^- + NO \rightarrow R'CHO + NO_2 \qquad (6.30)$$

$$R'\overset{+}{C}HOO^- + NO_2 \rightarrow R'CHO + NO_3 \qquad (6.31)$$

Eqns (6.29)–(6.31) indicate that the zwitterion may decompose to a reactive radical (which may enter the hydrocarbon chain oxidation of NO) and a formyl radical which results ultimately in the formation of formaldehyde. The zwitterion may also oxidize NO and NO_2 being reduced itself to an aldehyde. NO_3 is of some importance in photochemical smog because of its participation in reactions (6.32) and (6.33).

$$NO_3 + NO_2 \rightleftharpoons N_2O_5 \qquad (6.32)$$

$$N_2O_5 + H_2O \rightarrow 2HNO_3 \qquad (6.33)$$

High concentrations of inorganic nitrate on the atmospheric aerosol present during a photochemical smog may be explained by the production of nitric acid. An alternative source of NO_3 and hence nitric acid is the oxidation of NO_2 with ozone (reaction 6.34). This reaction is very much slower than the ozone oxidation of NO (reaction 6.6).

$$NO_2 + O_3 \rightarrow NO_3 + O_2 \qquad (6.34)$$

The afternoon decrease in the concentrations of hydrocarbons and O_3 may now be explained by the reactions considered in the preceding paragraphs. A corresponding oxidation of aldehydes by O_3 offers a similar explanation for the decrease in aldehyde concentration in the afternoon. The evening injection of hydrocarbons and NO by motor vehicles is removed almost immediately by the relatively high concentration of O_3 generated throughout the day. These chemical explanations can only be partial explanations as physical phenomena such as gaseous dispersion and interactions with solid surfaces have not been considered.

In attempting to explain the reactions occurring in photochemical smog, a motor vehicle pollutant gas produced in very high concentration has been omitted. This is CO which was, until recently, thought to be sufficiently unreactive to be ignored. Recent laboratory experiments suggest that this is not the case, so that reactions (6.35)–(6.41) should be considered.

$$NO + NO_2 + H_2O \rightarrow 2HNO_2 \qquad (6.35)$$

$$HNO_2 + h\nu \rightarrow OH\cdot + NO \qquad (6.36)$$

$$CO + OH\cdot \xrightarrow{O_2} CO_2 + HO_2\cdot \qquad (6.37)$$

$$HO_2\cdot + NO \rightarrow NO_2 + OH\cdot \qquad (6.38)$$

$$HO_2\cdot + NO_2 \rightarrow HNO_2 + O_2 \qquad (6.39)$$

$$OH\cdot + NO \rightarrow HNO_2 \qquad (6.40)$$

$$HO_2\cdot + HO_2\cdot \rightarrow H_2O_2 + O_2 \qquad (6.41)$$

These reactions (6.35)–(6.41) provide a further means of oxidizing NO to NO_2 without the participation of atomic oxygen or O_3. They also provide

a means for the rapid oxidation of CO to CO_2 which is of some importance
in the atmospheric chemistry of CO (see Chapter 7). The production of HNO_2
in reaction (6.35) and its subsequent photolysis in reaction (6.36) to produce
hydroxyl radicals is the initiation step of a chain reaction. The propagation
steps are reactions (6.37) and (6.38) which provide a means for the oxidation
of both CO and NO. Reactions (6.39) and (6.40) are termination steps that
produce HNO_2 which may then be used in chain initiation. The production
of H_2O_2 has been observed in photochemical smogs and reaction (6.41)
provides a mechanism for this. It is also another chain-termination reaction.
It is important to note that reactions (6.35) to (6.41) provide a pathway for
the oxidation of NO to NO_2 without the participation of hydrocarbons,
O_3, or atomic oxygen i.e. the photolytic NO oxidation cycle may be disrupted
in the absence of hydrocarbons. The rates of the reactions (6.4)–(6.6) involved
in the cyclic photolytic oxidation of NO to NO_2 are, however, much greater
than the rates of reactions (6.35)–(6.41), thus the photolytic oxidation is the
most important. The pathway through carbon monoxide is only supple-
mentary to it.

Another important pollutant gas that is usually found in photochemical
smog situations is SO_2. In Chapter 5 it has been pointed out that SO_2 arises
from many combustion processes and thus is usually emitted to the atmosphere
in which photochemical smog reactions may be occurring. An analysis of
the aerosol associated with photochemical smog indicates a high concentration of sulphate. This is thought to be derived by the oxidation of SO_2 to
H_2SO_4 in the presence of some of the constituents of photochemical smog.
Possible oxidizing agents include O_3, atomic oxygen, NO_2 and hydrocarbon
radicals.

In the absence of photodissociation reactions, NO_2 will oxidize SO_2 only
very slowly. When photodissociation of NO_2 occurs in the atmosphere the
atomic oxygen so produced can oxidize SO_2 rapidly. However, SO_2 must
compete for the atomic oxygen with other molecules, especially O_2 which is
present at a concentration of about 10^5 times greater than that of SO_2.
Oxidation with atomic oxygen is thus unlikely. The reaction between SO_2
and O_3 in a pure atmosphere is very slow indeed, but in an atmosphere
containing hydrocarbons, especially olefins the reaction is very rapid indeed.
The oxidation is supposed to be through the agency of peroxo zwitterions
(eqn 6.42).

$$\overset{+}{R}CHOO^- + SO_2 \rightarrow RCHO + SO_3 \qquad (6.42)$$

This reaction is analogous to the oxidation of NO by peroxo zwitterions
(eqn 6.30). SO_3 is rapidly hydrated with atmospheric water vapour to form a
sulphuric acid mist. This mist and the absorption of light by NO_2 in the visible
region of the solar spectrum, offer an explanation for the reduction in visibility
associated with photochemical smogs.

It is possible to postulate a photochemical smog situation arising from only natural compounds in the atmosphere. We have seen that the photocyclic oxidation of NO is basic to photochemical smog, and that NO can arise from the bacterial oxidation of NH_3. The largest natural source of hydrocarbons in the atmosphere is probably the organic volatiles emanating from plants. These include the terpenes that account for the blue haze seen over large forests. Most natural terpenes have the molecular formula $(C_5H_8)_n$ and are thus unsaturated. It is not surprising therefore, that laboratory experiments have shown that the terpenes, α- and β-pinene, are rapidly oxidized in an atmosphere containing irradiated NO_2.

Damaging effects of NO and NO_2

No cases of human poisoning from NO have been reported because of its relatively low toxicity. Although NO_2 has a higher toxicity to man even its milder effects such as mucous membrane irritation do not occur at atmospheric NO_2 concentrations. With an odour threshold in the range $1-3$ mg kg^{-1} it is seldom detected by the nose in photochemical smog episodes, where the maximum concentration is usually 0.3 mg kg^{-1}.

With vegetation, concentrations of NO_2 in excess of 2 mg kg^{-1} are known to cause leaf damage to sensitive plants. Such concentrations are seldom found in the atmosphere but NO_2 concentrations that are capable of inhibiting photosynthesis (about 0.6 mg kg^{-1}) in some plants are known. The mechanism for this inhibition is, at present, unknown. At lower concentrations plants absorb both NO and NO_2 from the atmosphere without damage. The uptake rate for NO_2 is about twenty times greater than that for NO and its absorption by plants could be significant in the total nitrogen nutrition of the plant.

Damage to materials in the environment by NO and NO_2 is slight. Because it is an oxidizing agent NO_2 may bleach certain dyes e.g. some dyestuffs are substantially faded by treatment for 6 days with 0.7 mg kg^{-1} NO_2 at 50 per cent relative humidity.

Damaging effects of PANs

A typical maximum concentration of PAN in a photochemical smog is 0.03 to 0.04 mg kg^{-1}. Acute toxicity is not found in humans at concentrations of this order but eye irritation occurs after a 12 minute exposure to 0.5 mg kg^{-1} of PAN. Cardio-pulmonary functions are altered at concentrations as low as 0.3 mg kg^{-1} of PAN.

Vegetation effects occur at considerably lower concentrations: sensitive plants may show leaf injuries at PAN concentrations as low as 0.01 mg kg^{-1} and even lower levels may contribute to early leaf fall. Damage usually takes the form of the collapse of young mature cells surrounding the air space in the stomata, hence young leaves are the first affected. Unfortunately, it must again be recorded that the biochemical basis for the damage is largely unknown.

Damaging effects of aldehyde

Maximum atmospheric aldehyde concentrations of up to $0.3\,\mathrm{mg\,kg^{-1}}$ have been recorded. Formaldehyde in concentrations of this order is known to have deleterious effects on human health. Eye irritation may be experienced at $0.2\,\mathrm{mg\,kg^{-1}}$ and at as low as $0.13\,\mathrm{mg\,kg^{-1}}$ dry and sore throats may be experienced. This latter effect may be related to the known inhibition of the respiratory cilia by $0.5\,\mathrm{mg\,kg^{-1}}$ of formaldehyde. Few data are available on the toxic effects of other organic components of photochemical smog.

7. Carbon monoxide

If CO_2 is disregarded, the most abundant gaseous pollutant in the atmosphere in the immediate vicinity of most towns is CO. This gas is produced by man by the incomplete combustion of carbonaceous fuels. The largest source of combustion-produced CO is the internal combustion engine. In busy city streets peak concentrations may be over 100 mg kg^{-1} while in tunnels carrying motor vehicles concentrations as high as 300 mg kg^{-1} are found.

Global emissions of CO from combustion sources are estimated at $2.6 \times 10^8 \text{ t y}^{-1}$, almost all of which is emitted in the Northern Hemisphere. For some time it was thought that this was the major source of atmospheric CO, and that natural sources were small and of little significance. Recent evidence now suggests that combustion CO constitutes only about 10 per cent of the total CO source, the remainder arising from two large natural sources—atmospheric reactions and emission from the oceans.

Atmospheric CO sources

Oxidation of CH_4 is probably the most important source of atmospheric CO. As almost all of the CH_4 in the atmosphere is produced by anaerobic decomposition of organic matter, this source of CO is of natural origin.

This oxidation is essentially in two steps—the oxidation of CH_4 by hydroxyl radicals (OH·) to formaldehyde via methyl (CH_3·), methylperoxo (CH_3O_2·) and methoxo (CH_3O·) radicals, followed by the photolysis of formaldehyde.

Methane oxidation is initiated by hydroxyl radicals hence a source of these species is necessary for the reaction to proceed. This is provided by the photochemical decomposition of ozone and the subsequent reaction of atomic oxygen with water vapour (reactions 7.1 and 7.2).

$$O_3 + h\nu \rightarrow O_2 + O· \tag{7.1}$$

$$·O + H_2O \rightarrow 2OH· \tag{7.2}$$

The importance of radiation-initiated reactions in the oxidation of CH_4 explains why this source of CO is often referred to as photochemical.

Oceanic CO sources

A second natural source of CO has been found to be the surface layers of the ocean. Samples of mid-ocean surface waters were found to contain up to ninety times the concentration of CO calculated from standard CO solubility data for the partial pressure of CO in the atmosphere immediately above the ocean surface. This implies that CO is produced in the ocean surface layer and that there is a flux of CO from the ocean to the atmosphere. The degree of supersaturation of CO in the surface waters is increased in the presence of sunlight suggesting two possible mechanisms for the CO production—photochemical oxidation of organic matter or biological oxidation by marine organisms. It appears that the latter source is the most likely.

The size of this marine source has been calculated assuming an average supersaturation of the ocean surface waters of twenty times the equilibrium value. This calculation indicates that the marine source and the combustion source are of approximately the same size in the Northern Hemisphere. The Southern Hemisphere, where combustion CO sources are small, is thus dominated by the natural production of CO.

Destruction of atmospheric CO

An atmospheric residence time for CO of about 0·2 years has been deduced from measurements made on atmospheric ^{14}CO. Cosmic ray neutrons produce ^{14}C atoms from nitrogen in the upper atmosphere. Almost all of these atoms react with oxygen producing ^{14}CO. A knowledge of the rate of production of ^{14}CO by this mechanism, together with a measurement of the abundance of ^{14}CO in the atmosphere, allowed the residence time to be determined.

A relatively rapid rate of CO destruction is required to satisfy such a short residence time. Measurements of the vertical concentration profile of CO show an approximately constant CO concentration throughout the troposphere with a sharp drop above the tropopause in the stratosphere (see Fig. 7.1).

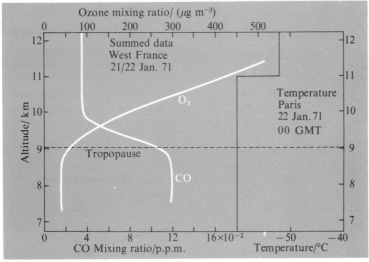

FIG 7.1. CO and O_3 concentration profiles from ascents over West France (from Seiler and Warneck 1972).

Following the rapid decrease a steady, but lower concentration is found through the remainder of the stratosphere. A very active destruction of CO thus occurs in the lower stratosphere.

The reaction thought most likely to account for the destruction of CO in the lower stratosphere is represented by eqn (7.3)

$$OH + CO \rightarrow CO_2 + H \tag{7.3}$$

The hydroxyl radicals arise largely from reactions (7.1) and (7.2). The constant CO concentration higher in the stratosphere is probably due to a source in this region whose rate of CO production is the same as the rate of CO oxidation in reaction (7.3). This source is probably the oxidation of methane which has diffused from the troposphere.

In the troposphere oxidation reaction (7.3) is also possible, especially in regions where photochemical smog is being actively formed (see reactions 6.34–6.40). Because much CO is emitted with other motor-vehicle exhaust gases, a considerable portion of the oxidation of exhaust-CO may occur rapidly. Within the troposphere as a whole, turbulent mixing with naturally-produced CO would make any observation of this oxidation very difficult.

At the surface of the earth CO may be destroyed by biological mechanisms. Soil fungi are known to absorb CO at an average rate of $2\ \mu g\ s^{-1}$ per square metre of soil. The total soil surface of the United States is sufficient to absorb three times the total combustion CO produced in the world. As combustion CO is produced close to the surface of the earth, soil fungal activity must be regarded as an important means of reducing the concentration of CO emitted to the atmosphere by man's activity.

Higher plants also have an ability to absorb CO, but the absorption rate is so low that it can be detected only by sensitive radiochemical techniques. The amount of CO taken up by plant leaves varies with plant species from zero up to about $2\ \mu g\ s^{-1}$ per m^2 of ground area. This suggests that plant leaves have an ability to absorb atmospheric CO comparable to that of soil fungi. Biological uptake of CO is thus very important in reducing tropospheric CO concentrations at the earth's surface.

Some aspects of the mechanism of CO destruction by higher plants are known. In the light, the predominant mechanism appears to be reduction of CO probably to 5-formyl-tetrahydrofolic acid which is a well-known biological carrier of groups containing one carbon atom. The reduced CO is fixed as the amino acid serine and thence enters normal protein and carbohydrate metabolism. In the dark, almost all of the CO taken up by plant leaves is oxidized to CO_2 and released again to the atmosphere. The rate of uptake of CO is the same in the light and dark and, as both light and dark processes lead to the destruction of CO, it can be taken that the plant leaves operate throughout the whole day destroying atmospheric CO at an approximately constant rate.

Toxicity of CO

The only known damaging effects of CO at the concentrations found in the atmosphere are related to animal respiratory systems based on haemoglobin as an oxygen-transporter. Haemoglobin is made up of four haem molecules, which are complexes of Fe^{II}, bound to one molecule of the protein globin. The haem molecule has the structure shown in Fig. 7.2. Magnetic

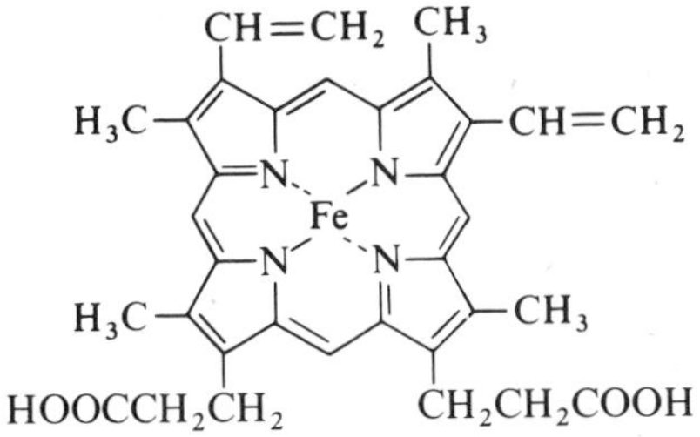

FIG. 7.2. Structure of the haem molecule.

susceptibility measurements have shown that Fe^{II} in haem is in a d^6 octahedral configuration with four unpaired electrons. The four nitrogen atoms of the organic chelate occupy four of the octahedral coordination positions in roughly the same plane. One of the positions perpendicular to the plane is occupied by coordination to the globin molecule, while the other position is available for coordination of (usually) oxygen gas. When oxygen gas is coordinated the molecule is known as oxyhaemoglobin.

The affinity of CO for the coordination position usually occupied by oxygen is about 200 times greater than the affinity of oxygen for the position. A relatively low partial pressure of CO is thus able to displace a considerable amount of oxygen from oxyhaemoglobin (HbO_2) to form the CO–haemoglobin complex known as carboxyhaemoglobin (HbCO), as shown in eqn (7.4).

$$HbO_2 + CO \rightleftharpoons HbCO + O_2 \qquad (7.4)$$

The transport of oxygen from the lungs to the tissues is thus impaired.

Reaction (7.4) is reversible, thus when the partial pressure of CO in the lungs is reduced, the reaction moves to the left and CO is released. The use of pure oxygen in the treatment of CO poisoning is based on this fundamental chemical principal.

Many individuals suffer from oxygen transport problems when the carboxyhaemoglobin content of their blood reaches 5 per cent. This is attained after exposure to 30 mg kg^{-1} CO for four hours or 120 mg kg^{-1} CO for one hour. It is of interest to note that cigarette smokers commonly have a carboxyhaemoglobin content in their blood of 5 to 10 per cent.

8. Minor pollutant gases

THE gases to be considered in this chapter are minor only in that their effects on the environment are overshadowed by the effects of the compounds discussed in earlier chapters. The compounds discussed are only a few of many that could have been chosen, however they have a common property in that they are all compounds that exist naturally in the atmosphere and have a man-made source.

The halogens

Fluorine is present in the unpolluted atmosphere in very low concentration in both gaseous and particulate form. It is most likely derived from sea-water aerosols formed by the bursting of bubbles in the oceans, especially in 'white caps'. The origin of the gaseous fraction is not clear. Oxidation of fluoride by ozone in the droplet phase, as has been suggested by some workers, is unlikely as the standard electrode potentials of O_3 and F^- are 2·07 and 2·87 V respectively.

Heavy industries contribute significant concentrations of fluorine to the atmosphere. The phosphate fertilizer industry releases fluorine largely as H_2SiF_6 from the action of H_2SO_4 on the fluorides in phosphate rock. Cryolite (Na_3AlF_6) is used as a flux with bauxite (Al_2O_3) in the aluminium industry. Fluorine released from this source appears as SiF_4, HF, or Na_3AlF_6.

Fluorine concentrations (gaseous plus particulate) can reach $5 \mu g\ m^{-3}$ in areas around industrial sources. These concentrations, as such, cause no immediate damage to plants or animals but fluorides are known to act as cumulative poisons to plants. Gaseous fluorides enter the leaves through the stomata and are transported to the edges of the leaves where they accumulate. With leaf concentrations in the range 50 to 200 mg kg^{-1} sensitive plants show leaf damage, however resistant species can tolerate up to 500 mg kg^{-1} of fluoride before being damaged.

Fluorine in the atmosphere does not have a direct effect on animals but grazing animals are indirectly affected. When atmospheric fluorine is accumulated in grass and forage crops the concentrations may be sufficiently high to cause fluorosis, which is characterized by bone damage. Safe fluoride levels in animal feed are about 50 mg kg^{-1}. Concentrations of this order are found in the vegetation about some large fluorine-emitting sources.

Chlorine also occurs in both gaseous and particulate forms in the unpolluted atmosphere. The source is largely sea-water droplets which may carry the chloride ion in solution or as a solid salt following evaporation. Both solid and gaseous chlorine in the atmosphere appear in the concentration range

$0.5-5$ $\mu g\,m^{-3}$. The nature of the gaseous form has not been determined but it is thought to be HCl and Cl_2. The former is thought to be formed in chloride-containing droplets by the action of H_2SO_4 on chlorides, (reaction 8.1), the H_2SO_4 being derived from the oxidation of SO_2.

$$2Cl^- + H_2SO_4 \rightarrow 2HCl + SO_4^{2-} \qquad (8.1)$$

Releases of free chlorine from industry are very unusual but a widespread pollution source exists in the form of motor vehicle emissions. These contain lead halide aerosols formed from 'anti-knock' compounds in the petrol. Photochemical decomposition of the halide produces chlorine atoms. These may participate in NO oxidation as shown in reactions (8.2)–(8.8).

$$Cl + O_2 + M \rightarrow ClO_2 + M \qquad (8.2)$$

$$ClO_2 + NO \rightarrow ClO + NO_2 \qquad (8.3)$$

$$ClO + NO \rightarrow Cl + NO_2 \qquad (8.4)$$

$$2ClO \rightarrow Cl + ClO_2 \qquad (8.5)$$

$$Cl + ClO_2 \rightarrow Cl_2 + O_2 \qquad (8.6)$$

$$2Cl + M \rightarrow Cl_2 + M \qquad (8.7)$$

$$Cl_2 + h\nu \rightarrow 2Cl \qquad (8.8)$$

Reactions (8.2)–(8.5) are propagation steps of a chain reaction oxidizing NO to NO_2 while reactions (8.6) and (8.7) are termination reactions. Reaction (8.8) is an initiation reaction that acts as an alternative to lead halide photo-decomposition.

An alternative man-made source of HCl in particular, is the combustion of chlorinated organic polymers, e.g. polyvinylchloride. The combustion of solid waste materials at rubbish dumps is a frequent source of this type.

Damage to human health by chlorine in the atmosphere has only been found in the high concentrations due to accidental spills. Plants, however, may be damaged by chlorine gas at concentrations as low as $0.2\,mg\,kg^{-1}$. At lower concentrations (about $0.1\,mg\,kg^{-1}$) this gas can cause partial closure of the leaf stomata. Even with this effect, which will reduce the uptake of chlorine by the plant leaf, it can be shown that vegetation uptake is an important means of reducing the concentration of atmospheric chlorine. Non-damaging concentrations of HF and HCl are similarly reduced when the atmosphere containing them passes over plant leaf surfaces.

Bromine has received little attention as a gas in the atmosphere. Like the other halogens its natural source is thought to be sea-salt droplets. It exists in the atmosphere in two forms—gaseous and particulate—with the former being about five times the concentration of the latter. The reasons for this observation are not yet understood.

The most important pollutant source of bromine in the atmosphere is the combustion of petrol containing lead 'anti-knock' compounds. The products of combustion include $PbClBr$, $NH_4Cl \cdot 2PbClBr$, and $2NH_4Cl \cdot PbClBr$. Photochemical decomposition of these lead halides result in the release of gaseous bromine. Some of this remains in the gas phase while the remainder may react with, or become absorbed on, solid or liquid atmospheric aerosols. Damage to vegetation by atmospheric concentrations of bromine has not been reported.

The concentrations of iodine found in the atmosphere fall in the range 0·01 to 10 $\mu g\,m^{-3}$ with an average of about 0·1 $\mu g\,m^{-3}$. As with the other halogens, both particulate and gaseous forms are found and these are thought to originate from the ocean. The concentration of gaseous iodine is several times greater than that of the particulate form. The gaseous iodine is generated at the surface of the ocean and is probably released directly from the ocean. Several mechanisms for the production of gaseous iodine have been proposed, including a photochemical reaction (eqn 8.9) and a chemical oxidation (eqn 8.10).

$$2I^- + \tfrac{1}{2}O_2 + H_2O + h\nu \rightarrow I_2 + 2OH^- \qquad (8.9)$$

$$2I^- + \tfrac{1}{2}O_2 + 2H^+ \rightarrow I_2 + H_2O \qquad (8.10)$$

Combustion of fossil fuels, which can contain up to 5 mg kg^{-1} of iodide, is a source of man-made iodine in the atmosphere. A radio-active isotope of iodine ^{131}I is released to the atmosphere as a result of nuclear fission. As ^{131}I is accumulated in the thyroid it has a high radiological toxicity. For this reason the chemistry of ^{131}I-iodine species in the atmosphere has been subjected to considerable study. The results obtained apply equally well to the naturally-occurring isotopes of iodine.

[131]I released to the atmosphere may be partially absorbed by the particulate matter in the atmosphere, the extent of the absorption varying with the concentration of particulate matter. In country districts with average smoke concentrations about 15 per cent is absorbed, while in air highly polluted with smoke up to 70 per cent may be absorbed. The gaseous forms of ^{131}I-iodine include I_2 gas and aliphatic iodides the most abundant of which is CH_3I. This latter is absorbed by particulate matter and solid surfaces to a much lesser extent than is I_2. Vegetation and soil have been found to be important surfaces for the removal of ^{131}I-iodine compounds from the atmosphere.

Hydrocarbons

In Chapter 6 the most important reactions of the hydrocarbon gases emitted to the atmosphere were considered in the discussion on photochemical smog. In this section the nature and origin of pollutant hydrocarbon gases will be considered together with the atmospheric chemistry of some of the most

TABLE 8.1

Average concentration of some hydrocarbons in urban atmospheres (from Grob and Altshuller et al.)

Hydrocarbon	Average concentration mg kg^{-1}
Methane	2·0
Ethane	0·05
Ethylene	0·03
Acetylene	0·03
n-Butane	0·03
Isopentane	0·02
Propane	0·02
Toluene	0·02
n-Pentane	0·02
m-Xylene	0·02
Isobutane	0·02
Propylene	0·01
Butenes	0·01

abundant compounds. In Table 8.1 are listed some of the hydrocarbons found in the urban atmosphere.

Methane appears to be present in an anomalously high concentration, but this is because of the considerable natural production of methane due to anaerobic bacterial decomposition of organic matter and emission from geothermal areas, coalfields, natural gas, and petroleum wells. This gives rise to a natural atmospheric methane concentration of about 1·4 mg kg^{-1}, and an estimated average life of methane in the atmosphere of up to 20 years.

About one half of the methane emission can be related to human activities which increase the turnover of organic matter at the surface of the earth. This could possibly be defined as pollutant methane. The production of methane in combustion processes, especially motor vehicle engines, is a smaller but more obvious pollutant source.

The ocean is almost in equilibrium with the partial pressure of methane in the atmosphere. There is a suggestion that the activity of micro-organisms in the surface layers of the ocean may be the origin of a small source of methane. The concentration of methane in the troposphere is virtually constant, but it falls off rapidly in the lower stratosphere. Oxidation by hydroxyl radicals to CO and water vapour is thought to cause this effect. Some concern has been expressed recently regarding the possible increased water vapour concentration in the stratosphere due to the oxidation of increasing methane emissions from the earth's surface. The water vapour may destroy the ozone equilibrium (see Chapter 4) by the production of hydroxyl radicals (reactions 7.1 and 7.2). A depletion in the stratospheric ozone concentration would give rise to an increase in the flux of ultraviolet light reaching the surface of the earth.

Of the hydrocarbons listed in Table 8.1 methane, ethane, propane, and to a lesser extent isobutane are derived largely from sources other than motor vehicle exhausts. Toluene and m-xylene have an exhaust origin as well as an industrial origin because of their importance as solvents in industry. The remainder are largely of motor vehicle exhaust origin. Methane, ethane, propane, and acetylene are of low reactivity in photochemical smog, while the olefins are of high activity.

Ethylene is the only hydrocarbon listed in Table 8.1 that is capable of damage to biological systems without further reaction and at atmospheric concentrations. The other hydrocarbons, because of their participation in reactions producing O_3, PAN etc. have an indirect damaging effect on biological systems. Ethylene is a plant growth hormone produced naturally at very low concentrations by many plants. Concentrations of 0.005 mg kg^{-1} of ethylene in the atmosphere can cause leaf damage to very sensitive plants, while less sensitive plants such as tomatoes can show growth retardation after exposure to 0.05 mg kg^{-1} for several weeks.

Ammonia and ammonium sulphate

Ammonia is released to the atmosphere in gaseous form. More than 80 per cent of the release is from natural sources especially the hydrolysis of urea from animal urine. No more than 20 per cent is released from the combustion of the nitrogen fraction of coal or from industrial processes using ammonia. Once in the atmosphere ammonia may remain as a gas or be found as the ammonium ion, largely in the form of ammonium sulphate. This latter is found in a crystalline form at relative humidities below 81 per cent and in the form of droplets at above 81 per cent relative humidity. The concentration of total ammonia (gas plus (NH_4SO_4)) in the atmosphere is about 5 μg m^{-3} and this remains relatively constant throughout the year.

The source of sulphate in atmospheric ammonium sulphate is largely SO_2. In Chapter 5 we have seen that SO_2 oxidation in the atmosphere may take place in the gas phase or the droplet phase. It is most likely that ammonium sulphate is formed in the droplet phase. When evaporation occurs ammonium sulphate particles are left suspended in the atmosphere. These are very effective condensation nuclei i.e. particles on which water vapour may attain the liquid state. This property enables atmospheric ammonium sulphate to exert an effect on visibility as very small water droplets markedly reduce visibility by light scattering. Calculations of this effect for a real situation show that when no condensation on ammonium sulphate particles has occurred (below 81 per cent relative humidity) the average visual range is 35 km while at higher humidity (90 per cent) when some condensation of water has taken place the average visual range is reduced to 24 km.

9. Indoor pollution

I N the foregoing chapters the impression may have been gained that exposure to air pollutants is an out-of-doors phenomenon. No real consideration has been given to indoor effects and this reflects the state of knowledge in this area. Very few data are available on the sources of indoor pollutants and of the reactions of these pollutants in a confined volume. In this chapter an attempt will be made to review the results of recent research in this area.

Indoor pollutant concentration

Studies with SO_2 have shown that the indoor concentration of SO_2 can be as low as 20 per cent of that prevailing outdoors. In one experiment a pulse of SO_2-polluted air was admitted to a room through a window. The SO_2 concentration in the room was continuously monitored and it was found that the concentration decreased with a first-order half-life of 40 to 60 minutes. Another experiment, carried out in a large test room coated with polyurethane resin, gave a half-life of 10 hours for the decrease in SO_2 concentration. Obviously the nature of the surfaces in the room has a significant effect on the behaviour of SO_2 in that room.

Data obtained for CO concentrations show that no significant decrease in CO concentration is found indoors compared with outdoors. This observation is consistent with the low capacity for absorption shown by CO. It also indicates the ease with which external air exchanges with indoor air.

Some measurements have been made of the indoor concentration of the solid atmospheric aerosol. A reduction of outdoor concentration to as low as 20 per cent is found indoors, the reduction being greatest in air-conditioned buildings. Concentrations ranged from 30 to 80 $\mu g\,m^{-3}$ of solid particulate material indoors in city buildings, both commercial and residential. Outdoor concentrations ranged from 80 to 190 $\mu g\,m^{-3}$. Analyses for Pb in the solid aerosol in buildings show that it follows the same concentration reduction as the total solid aerosol. Indoor concentrations in city buildings ranged from 0·2 to 2 $\mu g\,m^{-3}$ of Pb.

Interaction of pollutants with indoor materials

The reduction in the concentration of the solid aerosol indoors as compared with that outdoors, can be explained by the considerable capacity of solid aerosols to adhere to surfaces. This property has previously been mentioned in Chapter 2.

The comparable effect with SO_2 is due to its great capacity for sorption on surfaces. In recent years the capacity of many indoor materials to sorb SO_2 has been investigated. Some results are summarized in Table 9.1 where the

TABLE 9.1

Summary of average deposition velocities for SO_2 onto indoor surfaces

Surface	Velocity of deposition (cm sec^{-1})	Area of surface in a typical house (m^2)	Mass of SO_2 deposited in 1 hour from a pulse of 500 μgm^{-3} SO_2 (μgm)	Relative uptake based on unity for waxed linoleum
Wallpaper, PVC	0·003	100	5400	14
Wallpaper, cellulose	0·015	140	38 000	100
Timber, hardwood	0·048	—	—	—
Timber, softwood	0·024	5	14 000	37
Leather, upholstery	0.16	—	—	—
Linoleum, unwaxed	0·003	—	—	—
Linoleum, waxed	0·0006	35	380	1
Carpet, wool	0·021	50	19 000	50
Carpet, nylon	0·007	15	1900	5
Paint, gloss	0·020	60	22 000	58
Paint, emulsion	0·24	100	430 000	1130
Furnishing fabric,				
cotton	0·22	10	40 000	105
wool	0·29	10	52 000	137
artificial	0·026	20	9400	25

capacity of a given material to sorb SO_2 has been expressed as a velocity of deposition, v_g (see Chapter 5).

Also in Table 9.1 are the arbitrarily-selected surface areas of various materials in a 'typical house'. Use has been made of this parameter to show the likely distribution on indoor surfaces of SO_2 released within a house. The high v_g value for SO_2 deposition onto emulsion paint, together with the relatively high surface area of this material, makes it the most likely surface for the sorption of SO_2. In the 'typical house' these emulsion-painted surfaces have been assumed to be ceilings. It is of interest to note that about 40 per cent of the sulphur-35 SO_2 accidentally released in a laboratory was found on an emulsion-painted ceiling.

Upholstery leather also has a high capacity to sorb SO_2 but, as its surface area in a 'typical house' is small, it accounts for only a small portion of total SO_2 sorbed indoors. In the 'typical house' the second most important surface for the removal of SO_2 is furnishing fabric which accounts for about 15 per cent of the gaseous deposition. The natural fibres have a SO_2 deposition velocity of an order of magnitude greater than that of the artificial fibres. It is known that cellulosic fabrics undergo chemical changes, in the presence of atmospheric concentrations of SO_2, that lead to loss of strength probably because of H_2SO_4 hydrolysis.

Measurements of indoor and outdoor SO_2 concentrations in some Dutch homes have shown that indoor surfaces lose their ability to sorb SO_2 as time goes on. This was particularly obvious with painted surfaces and lime-washed surfaces. Indoor SO_2 concentrations in houses where the major absorbing surfaces were old thus approached the SO_2 concentrations prevailing outside.

Indoor pollution sources

Most pollutants measured within a building have a source outside that building, but some measurements have indicated that certain pollutants may arise from indoor sources. A particular example of such a source is a furnace operated to supply heating for a building. A poorly-maintained furnace may leak into the indoor air gases, such as CO and SO_2, as well as solid particulate material, depending upon the type of fuel used. In cases of this type the concentration gradient for CO and SO_2 may be from the building to the outside air.

Other indoor sources include gas stoves and gas heaters. These may give rise to their constituent organic gases (and CO in the case of coal gas) as well as a finite concentration of SO_2 from the combustion of sulphur-containing odourizers in the gas. Cooking also releases materials into the atmosphere, especially mists of cooking fats and oils. These are generally of a very small radius and remain in the air for some time before sedimentation. Their deleterious effect on clean surfaces about the cooking area is well known.

Cigarette smoking provides an indoor source of pollutants including CO and organic particulate material. Indoor sources of this type produce higher concentrations of pollutants than outdoors because of the limited volume of indoor air into which they are released.

Many people spend the major portion of the day indoors hence studies on the effects of air pollutants on human health should take into account the varying concentrations of pollutants to which people are exposed during the whole day.

References

Chapter 1

CHANDLER, T. J. (1967). *The air around us*, Aldus Books, London.
POGOSYAN, KH. P. (1965). *The air envelope of the earth*, Israel Program for Scientific Translations, Jerusalem.

Chapter 2

JUNGE, C. E. (1963). *Air chemistry and radioactivity* ch. 2. Academic Press, New York.
MILLER, M. S., FRIEDLANDER, S. K., and HIDY, G. M. (1972). *J. Colloid Interface Sci.* **39**, 165–176.
NOVAKOV, T., MUELLER, P. K., ALCOCER, A. E., and OTVOS, J. W. (1972). *J. Colloid Interface Sci.* **39**, 225–234.
PEIRSON, D. H., CAWSE, P. A., SALMON, L., and CAMBRAY, R. S. (1973). *Nature* **241**, 252–256.
STERN, A. C. (1968). *Air pollution* 2nd edn. vol. 1. Academic Press, New York.

Chapter 3

ATTIWILL, P. M. (1971). *Environ. Pollut.* **1**, 249.
DYER, A. J. (1972). *Proceedings of International Clean Air Conference, Melbourne, Australia, May* 1972, pp. 12–16.
GARRATT, J. R. and PEARMAN, G. I. (1972). *Proceedings of International Clean Air Conference, Melbourne, Australia, May* 1972, pp. 17–22.
HOOVER, T. E. and BERKSHIRE, D. C. (1969). *J. Geophys. Res.* **74**, 456.
JUNGE, C. E. (1963). *Air chemistry and radioactivity*, pp. 4–36. Academic Press, New York.
SAWYER, J. S. (1972). *Nature* **239**, 23.

Chapter 4

HILL, A. C. and LITTLEFIELD, N. (1969). *Environ. Sci. Technol.* **3**, 52–6.
JUNGE, C. E. (1963). *Air chemistry and radioactivity*, pp. 37–59. Academic Press, New York.

Chapter 5

EVANS, U. R. (1972). 'Mechanism of rusting under different conditions', *Br. Corros. J.* **7**, 10–14.
LISS, P. S. (1971). 'Exchange of SO_2 between the atmosphere and natural waters', *Nature* **233**, 327–9.
ROBINSON, E. and ROBBINS, R. C. (1970). 'Gaseous sulphur pollutants from urban and natural sources', *J. Air Pollut. Contr. Assn.* **20**, 233–5.
STERN, A. C. (ed.) (1968). *Air pollution* 2nd edn., vol. 1. Academic Press, New York.
URONE, P. and SCHROEDER, W. H. (1969). 'SO_2 in the Atmosphere', *Environ. Sci. and Technol.* **3**, 436–45.

Chapter 6

ALTSHULLER, A. P. and BUFALINI, J. J. (1971). *Environ. Sci. and Technol.* **5**, 39–64.
HECHT, T. A. and SEINFELD, J. H. (1972). *Environ. Sci. and Technol.* **6**, 47–57.
STERN, A. C. (1968). *Air pollution* 2nd edn., vol. 1 Academic Press, New York.

Chapter 7

BIDWELL, R. G. S. and FRASER, D. E. (1972). *Can. J. Bot.* **50**, 1435–9.
INMAN, R. E., INGERSOLL, R. B., and LEVY, E. A. (1971). *Science* **72**, 1229–31
JUNGE, C., SEILER, W., and WARNECK, P. (1971). *J. Geophys. Res.* **76**, 2866–79
WOFSY, S. C., McCONNELL, J. C., and McELROY, M. B. (1972). *J. Geophys. Res.* **77**, 4477–93.

Chapter 9

WILSON, M. J. G. (1968). *Proc. R. Soc. A.* **300**, 215–21.
YOCOM, J. E., CLINK, W. L., and COTE, W. A. (1971). *J. Air Pollut. Contr. Assn.* **21**, 251–9.

Index

1A	IIA	IIIA	IVA	VA	VIA	VIIA	VIII			IB	IIB	IIIB	IVB	VB	VIB	VIIB	O
₁H 1·008																	₂He 4·003
₃Li 6·941	₄Be 9·012											₅B 10·81	₆C 12·01	₇N 14·01	₈O 16·00	₉F 19·00	₁₀Ne 20·18
₁₁Na 22·99	₁₂Mg 24·31											₁₃Al 26·98	₁₄Si 28·09	₁₅P 30·97	₁₆S 32·06	₁₇Cl 35·45	₁₈Ar 39·95
₁₉K 39·10	₂₀Ca 40·08	₂₁Sc 44·96	₂₂Ti 47·90	₂₃V 50·94	₂₄Cr 52·00	₂₅Mn 54·94	₂₆Fe 55·85	₂₇Co 58·93	₂₈Ni 58·71	₂₉Cu 63·55	₃₀Zn 65·37	₃₁Ga 69·72	₃₂Ge 72·59	₃₃As 74·92	₃₄Se 78·96	₃₅Br 79·90	₃₆Kr 83·80
₃₇Rb 85·47	₃₈Sr 87·62	₃₉Y 88·91	₄₀Zr 91·22	₄₁Nb 92·91	₄₂Mo 95·94	₄₃Tc 98·91	₄₄Ru 101·1	₄₅Rh 102·9	₄₆Pd 106·4	₄₇Ag 107·9	₄₈Cd 112·4	₄₉In 114·8	₅₀Sn 118·7	₅₁Sb 121·8	₅₂Te 127·6	₅₃I 126·9	₅₄Xe 131·3
₅₅Cs 132·9	₅₆Ba 137·3	₅₇La 138·9	₇₂Hf 178·5	₇₃Ta 180·9	₇₄W 183·9	₇₅Re 186·2	₇₆Os 190·2	₇₇Ir 192·2	₇₈Pt 195·1	₇₉Au 197·0	₈₀Hg 200·6	₈₁Tl 204·4	₈₂Pb 207·2	₈₃Bi 209·0	₈₄Po (210)	₈₅At (210)	₈₆Rn (222)
₈₇Fr (223)	₈₈Ra 226·0	₈₉Ac (227)															

Lanthanides	₅₇La 138·9	₅₈Ce 140·1	₅₉Pr 140·9	₆₀Nd 144·2	₆₁Pm (147)	₆₂Sm 150·4	₆₃Eu 152·0	₆₄Gd 157·3	₆₅Tb 158·9	₆₆Dy 162·5	₆₇Ho 164·9	₆₈Er 167·3	₆₉Tm 168·9	₇₀Yb 173·0	₇₁Lu 175·0
Actinides	₈₉Ac (227)	₉₀Th 232·0	₉₁Pa 231·0	₉₂U 238·0	₉₃Np 237·0	₉₄Pu (242)	₉₅Am (243)	₉₆Cm (248)	₉₇Bk (247)	₉₈Cf (251)	₉₉Es (254)	₁₀₀Fm (253)	₁₀₁Md (256)	₁₀₂No (254)	₁₀₃Lw (257)

SI units

Physical quantity	Old unit	Value in SI units
energy	calorie (thermochemical)	4·184 J (joule)
	*electronvolt—eV	$1·602 \times 10^{-19}$ J
	*electronvolt per molecule	96·48 kJ mol^{-1}
	erg	10^{-7} J
	*wave number—cm^{-1}	$1·986 \times 10^{-23}$ J
entropy (S)	eu = cal g^{-1} °C^{-1}	4184 J kg^{-1} K^{-1}
force	dyne	10^{-5} N (newton)
pressure (P)	atmosphere	$1·013 \times 10^{5}$ Pa (pascal), or N m^{-2}
	torr = mmHg	133·3 Pa
dipole moment (μ)	debye—D	$3·334 \times 10^{-30}$ C m
magnetic flux density (H)	*gauss—G	10^{-4} T (tesla)
frequency (v)	cycle per second	1 Hz (hertz)
relative permittivity (ε)	dielectric constant	1
temperature (T)	*°C and °K	1 K (kelvin); 0 °C = 273·2 K

(* indicates permitted non-SI unit)

Multiples of the base units are illustrated by length

fraction	10^9	10^6	10^3	1	(10^{-2})	10^{-3}	10^{-6}	10^{-9}	(10^{-10})	10^{-12}
prefix	giga-	mega-	kilo-	metre	(centi-)	milli-	micro-	nano-	(*ångstrom)	pico-
unit	Gm	Mm	km	m	(cm)	mm	μm	nm	(*Å)	pm

The fundamental constants

Avogadro constant	L or N_A	$6·022 \times 10^{23}$ mol^{-1}
Bohr magneton	μ_B	$9·274 \times 10^{-24}$ J T^{-1}
Bohr radius	a_0	$5·292 \times 10^{-11}$ m
Boltzmann constant	k	$1·381 \times 10^{-23}$ J K^{-1}
charge of a proton	e	$1·602 \times 10^{-19}$ C
(charge of an electron = $-e$)		
Faraday constant	F	$9·649 \times 10^{4}$ C mol^{-1}
gas constant	R	$8·314$ J K^{-1} mol^{-1}
nuclear magneton	μ_N	$5·051 \times 10^{-27}$ J T^{-1}
permeability of a vacuum	μ_0	$4\pi \times 10^{-7}$ H m^{-1} or N A^{-2}
permittivity of a vacuum	ε_0	$8·854 \times 10^{-12}$ F m^{-1}
Planck constant	h	$6·626 \times 10^{-34}$ J s
(Planck constant)/2π	$\hbar$	$1·055 \times 10^{-34}$ J s
rest mass of electron	m_e	$9·110 \times 10^{-31}$ kg
rest mass of proton	m_p	$1·673 \times 10^{-27}$ kg
speed of light in a vacuum	c	$2·998 \times 10^{8}$ m s^{-1}

$\ln 10 = 2·303$ $\ln x = 2·303 \lg x$ $\lg e = 0·4343$ $\pi = 3·142$
$R \ln 10 = 19·14$ J K^{-1} mol^{-1} $RTF^{-1} \ln 10 = 59·16$ mV at 298·2 K